Bumblebee Keeper: a personal story of pollinator management
Nelson Pomeroy

ISBN: 978-1-914934-56-8
Text and graphics © Nelson Pomeroy

Published by Northern Bee Books 2023
Northern Bee Books, Scout Bottom Farm, Mytholmroyd,
Hebden Bridge HX7 5JS (UK).
www.northernbeebooks.co.uk
+44 (0) 1422 882751

Book design by www.SiPat.co.uk

Bumblebee Keeper
A personal story of pollinator management
Nelson Pomeroy

Author contact: nelsonpomeroy@gmail.com

Contents

Preface

Preface

Bumblebees are among the most admired insects – they are big, furry, approachable, noisy, and have the added benefit of being useful pollinators of crops and wild plants. This appealing combination has supported the publication of a profusion of books on the subject, especially during the last twenty years. A big influence on research and publications on bees in general has been an appreciation of their role in pollinating crops for human use, and the setbacks experienced by the honeybee industry during the early 2000s from the spread of the *Varroa* mite and the somewhat mysterious 'colony collapse disorder.' (The honeybee industry in many regions has recovered and worldwide beehive numbers are increasing.[1]) The surge in bumblebee research has also been stimulated by the existence of a unique new industry which has enabled researchers and crop growers to easily buy hives of bumblebees. The bumblebee industry which sprang into being at the end of the 1980s is well classified as 'unique' – where else are people raising livestock units, each with the social complexity of a wolf pack, on the scale of battery chickens?

I wrote this book as a 'technical memoir.' I had been fascinated by insects in general, and bumblebees in particular, from early childhood, and have made more than one unusual study and employment choice to keep working with them. As my career lurched about and overran previous projects, I accumulated research data, and ideas for practical techniques, that would be lost if, at my three-score years and ten, I did not publish them. Most of the book is a more or less linear account, from digging up and observing wild colonies in the hills of my childhood, to running a 'factory' rearing thousands of hives of bumblebees to serve the greenhouse tomato industry of New Zealand. Along the way I battled with the pollination 'numbers game' – the problem of working out how many bees work from a colony of bumblebees and how many you need for commercial-level pollination (Chapter 3).

I wanted to show a personal evolution of practical techniques. They are not intended as recipes for others to follow, so much as exemplars of a way of thinking and working – starting with what a bumblebee would want. The patent described in Chapter 4 might look absurdly complicated to those accustomed to rearing bumblebees in wooden boxes, but it lent itself to mass production and was exploited on the scale of many tens of thousands in Europe and North America.

Much has been written on preserving the biodiversity of pollinators and on bumblebee conservation, and it has been my friends' first line of conversation when have I told them about this writing project. In the last chapter I discuss examples from Australia, Britain and North America where bumblebees have been promoted or rejected. The issues are not simple. For example, some bumblebees are important for pollinating commercial food crops, but this advantage may not be the logical reason for protection of the endangered species. I suggest a more direct approach to conservation of rare bumblebees than the oft-promoted habitat improvements such as planting wildflowers.

Techniques and equipment designs are a big part of this book. I did not always work alone and cannot take sole credit for all of them. Where such equipment as electronic circuits were designed without my input I have tried to make that clear. And the patented 'starter cup' was developed over time with discussions with graduate students and technicians. I have tried to faithfully give credit to them in the Acknowledgements section, but apologise for omissions or inaccuracies. Much of the action was nearly four decades ago.

Terminology

Bee	To a specialist, a 'bee' is a member of the insect family Apidae. There are thousands of species of bees; most are small, solitary and do not store nectar. In popular English usage, it usually refers to the honeybee *Apis mellifera* which lives in large perennial colonies and is the basis of the apiculture industry. Here I reserve 'honeybee' for that species.
Bumblebee	A member of the bee genus *Bombus*, of which there are several hundred species worldwide, mostly associated with the cooler North temperate regions. Bumblebees typically form small seasonal colonies and store limited amounts of nectar. In biology the correct term is 'bumble bee' (two words) because it is a type of bee. Compare with 'butterfly' (one word) which is not a kind of fly and 'bluebottle' which is not a kind of bottle. However since the single word is in more common usage even in some scientific literature, I have used it here.
Colony	A group of bees living together.
Hive	A human-made container in which bees are placed.
Nest	The non-living material surrounding a colony
Comb	The living brood plus empty cocoons and wax honey pots.
Domicile	A human-made container placed in the field for voluntary occupation by a wild queen.

Currency

Unless otherwise specified the $ amounts are in New Zealand dollars.
NZ$1 = 0.65 US dollar, 0.6 Euro, 0.5 British Pound, approximately, over the period of the story.

Chapter 1

Bumblebees in the field

Bumblebees in the field

As a small child, I stroked the velvety backs of drowsy bumblebees, and when I was about eleven years old, I noticed bumblebees coming out of an old rat hole in a bank behind our farm-house. I found a spade and dug them up. At the end of the hole, there was a bundle of dead grass, out of which peeped bright yellow cocoons. I scooped up a few on the spade and took them home. While my parents and I ate lunch, we watched a damp, silvery-grey queen chew her way out of a cocoon. We all admired the musky-sweet scent of the bumblebee comb, and my parents did not comment on the gritty residue that fell onto the tablecloth from the subterranean exhibit. Thus began my ambition to 'keep' bumblebees.

Nature was my back yard. My father worked on, and later managed, a four-hundred hectare (thousand-acre) sheep and beef-cattle farm in South Taranaki on the margin of the 'hill country,' North Island, New Zealand.[1] I wandered freely, exploring the ponds, creeks and the bush, and we heard the shrieks of kiwi at night. After that experience of watching a queen emerge on the lunch table, I had so many questions about how bumblebees lived. I got lucky. During the summer holidays between high school and university, when I was seventeen, our pet terrier was digging into an old rat hole in a bank when bumblebees started coming out. I called him off, went back after dark, and dug away enough earth to expose the nest cavity against which I placed a sheet of glass before covering it again. I wanted to be able to watch the nest *in situ*. I could only revisit two days later, by which time cattle had trampled the disturbed earth. The bumblebee nest was still there but was tipped on its side, so I 'rescued' it by transferring it to a glass-topped box and took it home. A dark wardrobe seemed like a good place to observe them, and we had one against an outside wall. So I bored a hole through the house wall and inserted a copper pipe that fitted into the wooden box in the wardrobe so the bees could go outside to forage. My brother Eric told me I should have asked Dad first, but I don't recall any trouble over it. Holes can be patched. Forty years later, as a teacher, I also bored such a hole - in the metal window frame of my high school science laboratory. It was on the third level of a four-storey building, and I was keeping bumblebees there.

Figure 1.1 *Bombus ruderatus* colony in my first observation hive. **The nest was excavated from the field and kept in a simple glass-topped wooden box in a wardrobe. They foraged freely through a pipe and hole in the house wall. 1970.**

I noticed that in our district there were two types of bumblebees. One type had two similar yellow bands on the thorax, and the other type had a single band at the front and a much wider band further back on the abdomen. I later learned my two-stripers were the long-tongued species *Bombus ruderatus* and the one-stripers were the short-tongued *Bombus terrestris*. Both were introduced from Britain in 1888 for red clover pollination and were deliberately spread throughout the country.[2] There was another shipment in 1905, and it is assumed that the other two species, *B. hortorum* and *B. subterraneus*, which are also 'two stripers,' are the results of the later release. No one knows what other species were released but failed to establish. *B. ruderatus* has since become rare in Britain but occurs in fair numbers in parts of Europe and Morocco. It has had two common names: 'large garden bumblebee' and 'ruderal bumblebee,' neither of which I find helpful as it is about the same size and no more likely to be seen in gardens than the other common species in New Zealand, and 'ruderal' apparently comes from Latin for rubble and refers to weedy plants growing in newly disturbed habitats. The other species, *B. terrestris*, is commonly known as the buff-tailed bumblebee or large earth bumblebee. The British subspecies *B. terrestris audax* has light-brown colouring at the rear of the abdomen (at least on queens). There are several other regional subspecies of *B. terrestris* around continental Europe, with a variety of colourations.[3] I am happy to stick with the scientific names.

My excavated nest contained one full-sized and slightly ragged individual I assumed to be the queen and small ones I understood to be workers. I saw adults chewing their way out of cocoons, and they reminded me of newly hatched ducklings – with silvery-grey damp matted fur. Within a day, they coloured up to regular black with yellow stripes and a dull-white 'tail'. By then, I was aware that any insect with wings must be an adult, so little bumblebees were not babies. A few days after installing the colony, several big adults emerged - the same size as the queen. This puzzled me, as it was still mid-summer (24 January), and I thought new queens only appeared in autumn. I found later that this was a widespread misunderstanding, and at least one scientific paper had been written with this misapprehension and had led to a mistaken conclusion about 'overwintering' colonies in New Zealand.[4] It has since become clear that bumblebee colonies take only a few months to grow from a single queen to maturity (rearing new queens). In the warmer parts of New Zealand, many *B. terrestris* queens emerge from hibernation in late winter, and their colonies already reach maturity before Christmas. *B. ruderatus* queens emerge from hibernation in late spring and commonly reach maturity in mid-summer.

During my undergraduate years at Massey University in Palmerston North, I often kept bumblebee queens in my room and experimented with inducing them to start colonies. This did not lead to much success, but I was also learning from the library, particularly the 1912 classic by Sladen and Free & Butler's 1959 book.[5] I learned that long-tongued bumblebees were of great interest for pollinating red clover, and that bumblebee colonies were very small – seldom reaching more than a few hundred workers compared to the tens of thousands in a honeybee colony. I wanted to continue to graduate work and research a thesis on bumblebees. My study species was going to be *B. ruderatus* which was abundant on our home farm. Foxgloves (*Digitalis purpurea*) massively bloomed in November, coinciding with the nest establishment by the queens. For several years, foxgloves on the farm had been controlled by the herbicide 2,4-D as an additive in superphosphate fertiliser. The farm applied about fifty tonnes of it by aircraft on ground too rough or steep for trucks. At the time,

2,4-D was considered low toxicity, and as it selectively kills broadleaf plants but not grasses, it was used extensively against weeds on grassland farms. But when fertiliser was dropped from aircraft, it often drifted. Thus homestead gardens and commercial crops were sometimes dusted with the herbicide, which resulted in so many damage claims against the topdressing companies that they eventually stopped applying that formulation. The foxgloves bounced back in the season my project began. Later in the summer, thistles and roadside red clover provided additional forage.

Figure 1.2 Wild foxglove flowers. The hillside is on my brother's farm, where domiciles were placed.

I wanted to study the way bumblebee colonies grew and why they stopped growing long before the onset of winter weather. But first, I needed to obtain colonies to observe. I discovered a few colonies of both species on the farm by chance. As Free and Butler advised, you need to observe the movements of every bumblebee you encounter. Almost all will be foraging – travelling between flowers – but occasionally, one will be seen descending to, or arising from, a place where there are no flowers, and thus probably a nest site.

I needed a more efficient method of finding nests. I tried tethering a small feather to a worker with fine thread to slow it down so I could follow it, but the bee usually flew to the nearest shrub and proceeded to preen itself in a vain attempt to remove the burden. So I tried a different approach. I reasoned that foragers would go home when they had acquired a full load of nectar, so if I could give them enough nectar on a single flower to fill them up, perhaps they would fly straight home. That should give me the direction to their nest, at least. I sometimes achieved this with the broad tubular blooms of foxglove. I would wait until a worker went inside the flower, after which I could rush forward unseen, insert a syringe needle into the base of the flower, and squirt in a few drops of diluted honey. Sometimes the liquid ran out of the flower, sometimes the bee was scared away, and sometimes she seemed to miss the extra 'nectar' and leave after the typical visit time of a few seconds. But sometimes, she spent nearly a minute inside the flower drinking the unexpected bounty. When she emerged, she would often pause at the lip of the flower and perform a few preening movements, followed by a short flight to a nearby flower head or thistle. There, she would preen some more before taking flight and circling the area a few times to memorise the location of the new bonanza and fly away in a straight line, often towards one of the many bush-clad gullies. By going to the bush edge and waiting, I'd sometimes see another worker arriving, and I'd work my way into the bush to their nest location. I found one nest by triangulation: the intersection of flight lines of two nectar-loaded workers.

Nectar-loading foxgloves required finesse, but Scotch thistles (*Cirsium vulgare*) that flowered later in the season were much more straightforward. The big upright brush-topped flowers would retain diluted honey that was dripped from above. Old workers with ragged wings could overload on the super-meal: they would try to fly away only to sink to the ground within a few metres. The poor things would climb the nearest grass stalk and try again with the same result. I resorted to divesting them of part of their load by (with gloved hands) pressing their abdomen shorter, which forced some regurgitation. Relieved of their overloading, they could fly again, but I assumed the procedure would have been so traumatic that I didn't trust that their flight path would be a useful guide to the home direction.

Most of the twelve *B. ruderatus* nest sites I examined over the years were in the underground cavities of abandoned mammal nests. Judging by tunnel widths, there were both mouse and rat holes, and one was at the end of a three-metre rabbit hole. Several had wide-open entrances walled or roofed by large tree roots.[6] This contrasted with nests of *B. terrestris* I have found over the years, which often have narrow, concealed entrances, which at the early stage are additionally camouflaged by the founding queen dragging in surrounding debris. I discovered one *B. ruderatus* nest only by hearing the queen buzz as I bumped some shelves in my parents' garage. One metre from the door (which had a cat-sized cut-out at the bottom), on the bottom shelf of a storage unit, there was a large clear plastic bag in which I had once collected hay for bumblebee experiments. A mouse had apparently nested in the bag, as there was a ragged, mouse-sized hole at the base. The queen had found her way in through the hole and was nesting in the hay. This seemed such an unlikely place for her to find by visual searching (the garage was windowless) that mouse scent seemed to be a likely attractant, even though experiments by myself and others do not seem to have succeeded in attracting bumblebees to nest sites using mouse scent.

Excavating the wild colonies could be quite an adventure. I always did it after dark when all the foragers should be at home.[7] Bumblebees do not normally fly in the dark, and do not see red light very well, so a red-filtered torch was useful.[8] At least *B. ruderatus*, unlike *B. terrestris*, seldom attacked me during the operation. Several nests were behind thick tree roots, and a chainsaw was sometimes required. Free and Butler's book gives sound advice on excavating bumblebee nests.[9] Unless a bumblebee colony is very small or docile, the workers must be collected one by one as soon as they emerge in response to the disturbance. Ideally, when the nest is reached, most of the more aggressive workers will already be removed. I used to capture them in small glass tubes but later changed to using tweezers. Grabbed by a leg or wing, they are seldom harmed. They generally do not escape if deposited in a clean milk bottle or jar with a small hole in the lid. After the brood comb is transferred to an observation hive (or whatever accommodation is used), the workers can be dumped back into the nest.

Field domiciles

There has been a long history of placing artificial nesting sites in the field for voluntary occupation by colony-starting queens. The use of field domiciles is sometimes called 'trap nesting.' It is a way of intercepting wild queens if they choose to nest in the provided site. Much of the previous work in New Zealand had been done near Christchurch on South Island, where *B. hortorum* was the common long-tongued species, and it regularly occupied wooden box domiciles on the ground surface.[10] Initially thinking *B. ruderatus* might occupy similar sites, I made about a hundred small boxes of foam polystyrene (internal size 16 x 16 x 10 cm), reinforced with pasted newspaper and painted for weather resistance.

I tested the polystyrene box domiciles both around the farm and in a flight room. Like many of the assets exploited in my early research, the flight room was fortuitously available at the right time. My old primary school had been built on a corner of the farm. I was one of the last pupils to be taught there, in 1966, and it had been sold back to the farm. My father had removed the main windows for access, and had used the space to store hay, but it was vacant that spring. It was a single high-ceilinged room 5.2 x 6.4 m (17 x 21 feet – I remember measuring it in arithmetic classes), with a small porch. I re-covered the window spaces with polythene sheet, and whitewashed the walls for uniformity. Although *B. ruderatus* became my study species, I initially wanted to test the nesting habits of all three North Island bumblebee species. I hoped to use any resulting colonies for study. I collected ten queens each of *B. ruderatus* and *B. terrestris* on the farm and ten *B. hortorum* in the gardens at Massey University. *B. hortorum* had been introduced from the South Island to Palmerston North by Lou Gurr ten years previously and had become quite common there.[11] The queens were all released into the school room with nest boxes attached to walls at varying heights. There were four nesting materials:

▷ Mouse nests made of a mixture of hay and upholsterers' cotton.[12]
▷ Plain upholsterers' cotton.
▷ Dried fine moss.
▷ Mixture of chopped hay and sisal fibre (from old hay-baling twine).

Figure 1.3 Experimental flight room. Interior of my disused primary school, preceding my MSc thesis project, 1974. There are nest boxes at different heights, some of which were occupied by *B. hortorum* queens. A honey-water feeder is visible in the foreground.

The queens readily consumed honey-water from artificial feeders but needed real flowers for pollen, and this was a challenge. In the beginning I transplanted columbine plants (*Aquilegia vulgaris*) from our home garden, plus a few foxglove plants. Fortunately, the Botany and Zoology department had hundreds of bucket-sized plastic pots, which were the outer shells of hair salon dryer hoods, salvaged slightly smoke-damaged after a local factory fire. Foxgloves are big plants to transplant, and as the season progressed and more came into bloom, I collected flower stalks and placed them in the room in large beer bottles of water.

Figure 1.4 The three species in the flight room.
From left to right: *B. ruderatus, B. hortorum, B terrestris.*

I observed queens of all three species exhibiting nest-searching behaviour. They would fly slowly near floor level, pausing frequently and crawling among the debris. Although all three species occasionally went inside the boxes on the walls, only *B. hortorum* queens initiated colonies in them. The four occupied boxes were at various levels above the floor, and all contained plain upholsterers' cotton. This was by far the finest fibre and suggested that the texture, and hence thermal insulation, of the nest material was more important than the odour of mice. The failure of *B. ruderatus* and *B. terrestris* queens to use the (tunnel-less) nest boxes was consistent with my subsequent finding of most colonies of those species in underground mammal burrows. I had also placed nearly a hundred polystyrene box domiciles around the farm on the ground surface under scrub and ferns, but none were occupied.

The non-occupation of polystyrene boxes, and the places I had found wild colonies, suggested I would need underground domiciles to obtain colonies of *B. ruderatus*. The challenge was to create a tunnel leading to an underground nesting cavity. A trained rat seemed impractical. Sladen solved the problem by digging a pit, covering it with a tight lid, and punching a tunnel diagonally through the soil to emerge near the bottom of the pit. He used a metal rod with a pointed bulbous tip that created a tunnel by a similar principle as the tractor-drawn 'mole plough', which drags a bullet-shaped bulb through the soil at the end of a vertical blade. Sladen's system worked quite well, but the tunnels needed frequent clearing.[13]

Since then, most designs of underground domiciles have used a plastic tube as the tunnel. I had already seen how quickly plastic tube tunnels became wet with faeces (from observation hives; more on this in Chapter 2) and felt that at least the floor of the tunnel and nest cavity should be bare earth. But there would need to be a rigid structure to hold up the roof of the tunnel and the nesting cavity. Ideally, this would be porous, but I could not find low-cost earthenware pots of a suitable size. So I resorted to plastic – the same hair-dryer hoods I already mentioned for potting foxgloves. As the colonies were intended to be taken away at a young stage, i.e. before too much moisture was likely to accumulate, I felt the plastic pots were a suitable compromise. The tunnel was formed by digging a trench and laying a length of ribbed plastic drainage pipe in the bottom. I slit the 10 cm (4 in) diameter pipe into thirds lengthwise, which formed a tunnel with a very flexible arched roof and an earth floor. (For later versions, I was able to get 65 mm (2.5 in) diameter pipe which I slit in half). Several wild nests were in burrows with thick tree roots forming a roof over the entrance. To mimic this and to stabilise the entrance, I faced the entrance with a rough-split slab of wood with a truncated, inverted, 'V-shaped cut-out (Figure 1.6).

Upholsterers' cotton was an excellent nest material, but I also knew of bumblebees nesting in old jute carpet underfelt. In the domiciles, I placed a couple of fistfuls of upholsterers' cotton wrapped in a loop of underfelt. The cylindrical bundle was placed on a bed of twigs to reduce moisture absorption from the soil.

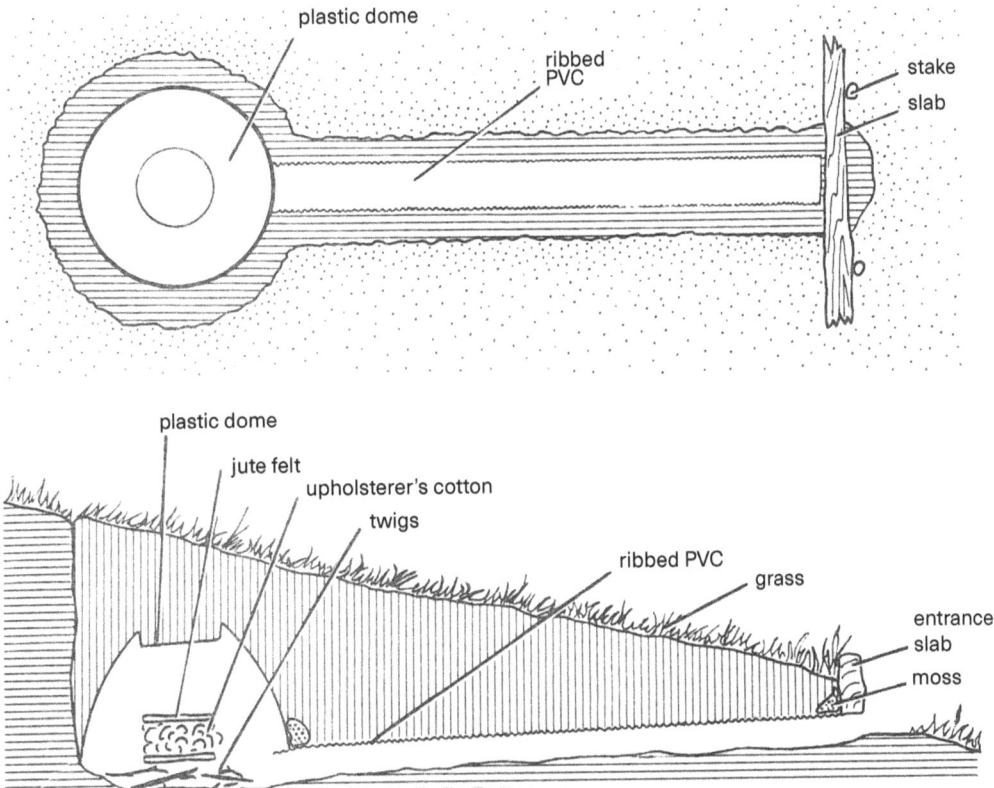

Figure 1.5 Drawing of underground domicile. **View from above before burying, and vertical section.**

Figure 1.6 Photo of underground domicile entrance. **Faced with a wooden slab held by two stakes, to prevent earth caving in and blocking the entrance, and to make it look like a gap in tree roots.**

Figure 1.7 Drawing of brick
domicile. **Showing brick layout
only. There was a lid of two
more bricks holding down either
a wad of newspaper or a thin
slab of foam polystyrene.**

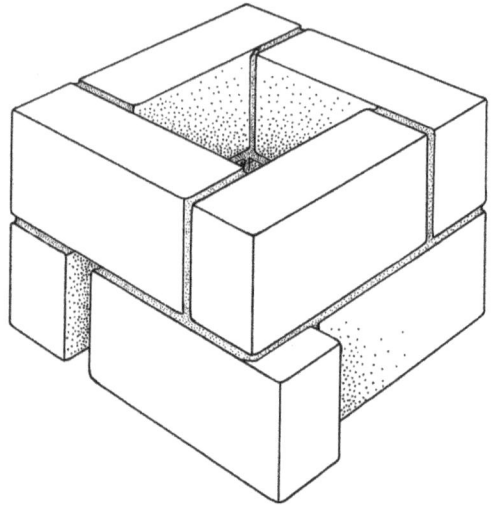

The domiciles needed to be installed before the *B. ruderatus* queens were nest-searching (November), but I did not want them occupied by earlier-emerging *B. terrestris* queens, so I covered the entrance holes with wire screen, thumb-tacked to the wooden slabs, and unblocked them in mid-October. I installed fifty of them, which entailed days of digging. Fortunately, the soils in the area were free of rock and gravel. I chose sloping sites, often close to trees or scrub and avoided low hollows that might become wet after rain.

Coincidentally that year, a supply of used red bricks became available from dismantling the chimney of the old school, and I decided to use them to make 'earthy' surface domiciles. I made twenty-three of them (Figure 1.7). Four bricks formed a square, with a second layer to increase the height, and a lid of polystyrene foam, held down by two more bricks, formed a compact cavity. One of the lower bricks was offset by about two centimetres creating a vertical slot as an entrance.

One could not just lift a lid to inspect the underground domiciles for occupancy. Nor would it have been practical to stand at the entrance to watch for forager traffic as you can wait many minutes to see a worker pass in or out of a tiny colony, and when there is just the queen, she may stay inside for hours at a time. So I exploited bumblebees' sensitivity to vibration: I stamped on the ground over the nest cavity while listening for the responding angry buzz through a long plastic tube with one end in my ear and the other end down the tunnel (functioning like a stethoscope). Alternatively, I could use the tube to blow down the tunnel, which had a similarly disturbing effect, and then listen through it, although my ear wax gave the tube a bitter taste. After the queens' first brood of workers emerged, their buzz added a higher-pitched harmony which sounded different. I preferred to do these checks at night to avoid overlooking any queens out foraging, but it involved some tricky clambering in the dark. Figure 1.2 shows typical terrain.

Of the forty-five domiciles that I could find again or were not buried by minor land-slips, forty-two (92%) were occupied by a queen at some stage, and in twenty-five (55%) I recognised workers; the majority were B. *ruderatus* (77% of those that were identified to species). Ninety-two per cent may be a world record for domicile occupation. However, I have since found that a high proportion of incipient colonies in the field never survive even to the emergence of any workers. So other field domicile trials may have had equally high initial occupation rates by queens, which may have been overlooked if they were not inspected within the first couple of weeks. Six (26%) of the brick domiciles were occupied by B. *ruderatus* queens.

Although the fully-underground domiciles seemed to suit B. *ruderatus* queens very well, removing the colonies was no mean task. They needed to be dug up, and it needed to be done after dark to avoid missing foraging workers. However, I did not see this as a problem during that early stage. My first aim was to find what was an attractive nest site, and worry later about making it economical and user-friendly. That season I only needed around ten colonies for my research, so I did not need to excavate all the occupied domiciles.

The following year (1975), I only needed around six colonies and decided to revert to surface domiciles, especially since the brick ones had been moderately successful. Thus began adventures with concrete. I quite like concrete. The raw material is cheap, and the finished product is durable. Most concrete products need to be non-porous to be weathertight or to contain water, but if made with a porous aggregate (the gravelly part added to cement), it remains porous rather like soil. Pumice, readily available in volcanic New Zealand and sold as a potting additive, was the perfect material. I used a pebbly grade of pumice (2 – 7 mm particle size) mixed with Portland cement in a 6:1 ratio. One of the reasons manufactured concrete products are much more expensive than the raw material is the need for moulds or form-work. Making multiple moulds to cast bumblebee domiciles would have been prohibitive, but I found a way to use a single mould repeatedly without waiting for the concrete to set. (I get annoyed by talk of concrete 'drying'. It *sets*, which is a chemical reaction and needs wetness.) A stiff pumice mix, firmly rammed tight, will stand up by itself. I used two sheet-steel concentric cylinders lined with oiled corrugated cardboard for a mould. After the concrete mix was placed and rammed down, the metal sleeves could be slid off with the cardboard staying attached to the concrete. Having tested different wall thicknesses (25, 50, 75 mm (1, 2, and 3 in)) and measuring maximum/minimum temperatures, I concluded that 50 mm thick walls were a good compromise between temperature stability and material economy.

The lids were shallow cones of bitumen-impregnated fibrous sheet (Malthoid®) partially filled with pumice concrete. The wide (one metre (yd)) sheet is sold as 'roofing felt' in some countries. In New Zealand, it is now only available in a narrow version for damp-coursing under timber frames (Figure 1.8).

Even the lightweight pumice produced a relatively heavy domicile. I made them at Massey University and transported them to the farm in the departmental Land Rover. I parked the vehicle in my father's garage before unloading the twenty-five domiciles, after which the hard-top Land Rover with de-compressed springs would not fit under the door. I needed to let air out of the tyres to extract the vehicle from the garage. Fortunately, the farm had a motorised pump to re-inflate the tyres when outside.

These concrete domiciles were intended to improve internal space, breathability and rain-proofness, compared to the brick type. I placed out twenty-five each of the brick and concrete types that season. The occupancy was less than the first year, with only four *B. ruderatus* in the brick domiciles and one of each species in the concrete type. I wanted more colonies than that for my research, but fortunately, I found that some of the old plastic-domed underground domiciles from the previous year were being reoccupied, even though I had not renewed the nest material, which was damp and rotting. So with some excavating of the buried domiciles, I was able to get six *B. ruderatus* colonies for transfer to the observation hives at the University soon after the first workers had emerged.

Underground nest sites were very attractive, and near the end of my thesis work, I had one more try to make a more user-friendly, bumblebee-attractive field domicile. I partially buried thirteen concrete domiciles and connected them to the surface with tunnels of a similar style to what I'd used for the fully underground ones. These worked quite well: over 60% attracted *B. ruderatus* queens, and the lids made inspection and colony removal straightforward.

That was as far as I went with field domiciles during my MSc thesis. Years later, I wanted a much larger number of *B. ruderatus* colonies for a trial on red clover pollination and needed to make hundreds of domiciles (details of the trial in Chapter 3). But casting box or cylinder shapes in large numbers did not seem practical. Instead, I made flat slabs with a tab on one side and cut-out on the other that could slot together into a square box. The idea was that when partly buried, the soil pressure would hold the sides together. A separately cast lid was a flat square with a foam polystyrene insert for insulation and a small lip around the perimeter to prevent it from sliding sideways. The walls were pumice and cement; the lid used pea-gravel and cement. All the components were made from a dryish mix that could be compressed and immediately de-moulded. Wall pieces were cast in a stack of four, separated by layers of newspaper. I made 350 of them in the garage of my father, now retired from the farm and living in the nearby small town of Waverley. The farm was now run by the previous owner's grandson, who gave me the run of the farm for the project.

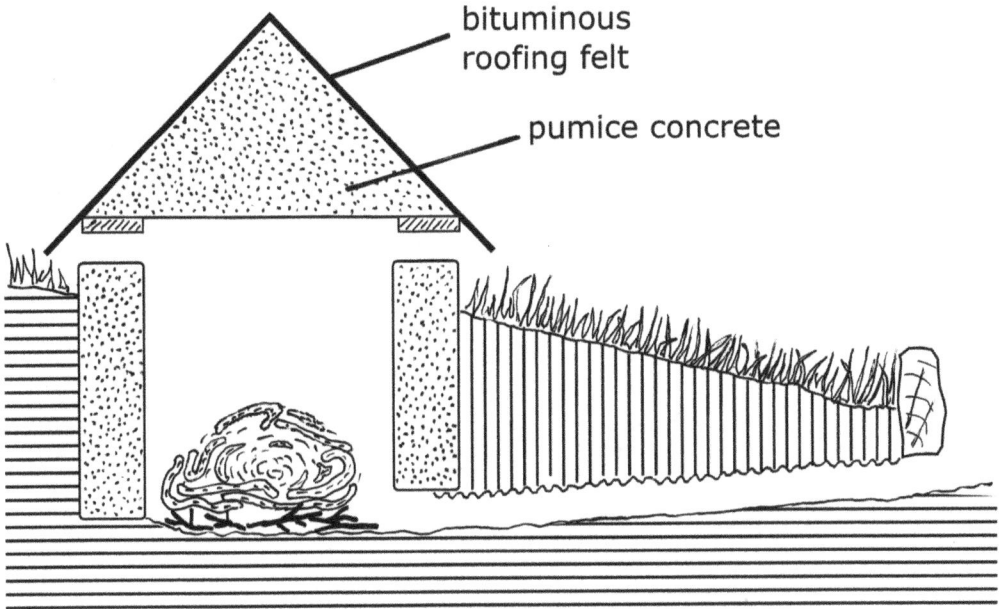

Figure 1.8 Drawing of semi-underground concrete domicile. **Unlabelled parts are the same as for the fully undergound domicile (Figure 1.5).**

Some domiciles were also placed on my brother Eric's farm 15 km further inland to the NW and approximately 350m altitude.

All domiciles were placed on livestock farms, and a few were damaged or had the lids displaced, presumably by cattle. They were inspected several times, and 6% were occupied by *B. terrestris*, the queens and brood of which were removed. Most of these were subsequently occupied by *B. ruderatus*. Of the 350 placed out, 70% showed occupation by queens, and 58% produced workers of *B. ruderatus*. Nearly 150 colonies were removed for a pollination trial described in Chapter 3.

It is difficult to assess why some other domicile designs have been relatively unsuccessful.[14] I suggest the visual effect of the entrance is important. Plastic tubes, unless constrained in some sort of entrance block, tend to project out of the ground in a very un-tunnel-looking way. When I was working with school students many years later, we made similar semi-underground domiciles using flexible ribbed electrical conduit tube as the tunnel. The entrance was cast into a solid concrete block of about 12 cm (5 in) cube. These were quite successful in the school grounds. The nest material is almost certainly essential. Following from the trial in the old school building, I suggest further research could be done in a large screen cage where various entrances and nest materials could be offered to confined queens in a relatively natural arena.

Figure 1.9 Flat-slab domicile parts. **The four walls were slotted together with the tab fitting in the cut-out. The walls were held in place by compressing the refilled soil.**

Figure 1.10 Flat-slab domicile partially installed. **The yellow tunnel was covered with the removed turf.**

A fortuitous meeting

The work with field domiciles was a means of obtaining colonies for further study: that is described in Chapter 2. I wanted to observe details of how colonies grew, to try and understand why they stopped growing. But I also thought they could become part of a commercial pollination industry for red clover seed production, and it would be helpful to get some industry support. One of the biggest seed companies in New Zealand at the time, Wrightson NMA Ltd., was a subsidiary of Challenge Corporation which happened to be having a conference at Massey. So I put on some shoes, combed my hair and bowled up to one of the suited gentlemen with the introduction that I was doing research that might benefit the seed industry. Not too dismissively, he gave me his card and said to write to him about it, so I did. He wrote back with encouraging words and a cheque for $100. Even then, it was a token sum as a research grant, but it caused some trouble with the University as mere students were not expected to solicit funds. However, the senior heads smiled benignly and said I could keep it. This connection sowed the seed for a major collaboration a few years later.

Nelson Pomeroy **Bumblebee Keeper**

Chapter 2

Bumblebees in the laboratory

Bumblebees in the laboratory

The field domiciles I described in the previous chapter were a 'means to an end' of obtaining colonies of *Bombus ruderatus* so I could study their growth and decline in detail. At this time (mid 1970s), pollination of red clover seemed to be the main potential used for 'bumblebee keeping', and colony size was obviously a factor in any economic consideration.

Before settling into the MSc thesis work at Massey University, I had 'kept' bumblebees in two different situations: whole active colonies dug up from the field, and confining individual queens to start colonies. Colonies from the field survived quite well in shallow, glass-topped wooden boxes, although they tended to cover the comb with a canopy of wax, and faecal moisture built up in the corners. Starting colonies from captive queens is a topic I deal with in more detail in Chapter 5. After I had given up on providing nest material (which always hid the action I wanted to see), I used electric heating. I started a few colonies, and their containers became the prototype observation hives. I was not aware at the time that Shôichi Sakagami, working in Brazil, had used electrical heating in an observation hive for stingless bees (Meliponini), and later for the bumble bee *Bombus atratus*.[1]

Observation hives

Early designs

My first 'hive' without nesting material was a cylindrical .303 ammunition box with extra foil insulation, a car tail-light bulb and a fish-tank thermostat (Figure 2.1).

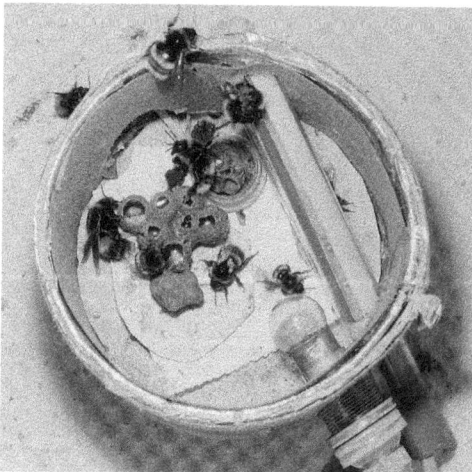

Figure 2.1 My first heated hive, in a cylindrical box. A 12v, 5W light bulb provides heat and is controlled by a fish-tank thermostat (switched by a bimetallic strip and encased in a waterproof glass tube). This small colony was started by a captive queen supplemented with captured workers. Honey pots in the centre, a larval clump below them and a plastic cap with pollen above. Workers 'forage' in a larger outer wooden box for diluted honey.

Bumblebee warmth

Although we think of insects as 'cold-blooded', all living things generate some metabolic heat. Active insects create more body heat, on a gram-for-gram basis than mammals or birds.[2] Most insects are so small, the heat is lost from their bodies and only big insects actually get much warmer than their surroundings. The large size of bumblebees gives them a sufficiently low surface-area : volume ratio, aided by their fur, that their internal temperature is able to be elevated. This happens inevitably with the heat generated by the wing muscles in flight, but they can also generate heat when unmoving. The heat is generated in the flight muscles and it was thought that the muscles 'shivered' with the wings disengaged, but later there was no evidence found for muscle shivering. Regardless of the precise mechanism, bumblebee queens (and later workers) are able to warm their brood just as a birds incubate their eggs. Indeed, a queen brooding on her initial brood clump behaves much like a broody chicken: she will make an agitated buzz when disturbed and may wave a leg, but can be reluctant to move.

The initial nest cavity is only about the size of a walnut, and the snug-fitting insulation enables a queen to warm her brood to around 30°C (86°F). If a captive queen is surrounded by close fitting insulation like cotton, she can warm her brood easily but the warmth retention diminishes rapidly if the insulated space is much larger, due to the larger surface area. I found early on that however well I insulated I made a practical-sized bumblebee hive, the bees' body heat could not warm the space inside very much because of the big surface for heat loss. Only a close-fitting envelope of insulation would work, but that would mean covering up what I wanted to observe, or at least designing insulation that could be incrementally expanded. Hence the decision to leave the brood comb exposed for observation and replace the natural benefit of close-fitting insulation with artificial heating.

As I became more successful at starting and maintaining colonies, I made a larger hive with a warmed, insulated 'nest' chamber (Figure 2.2).

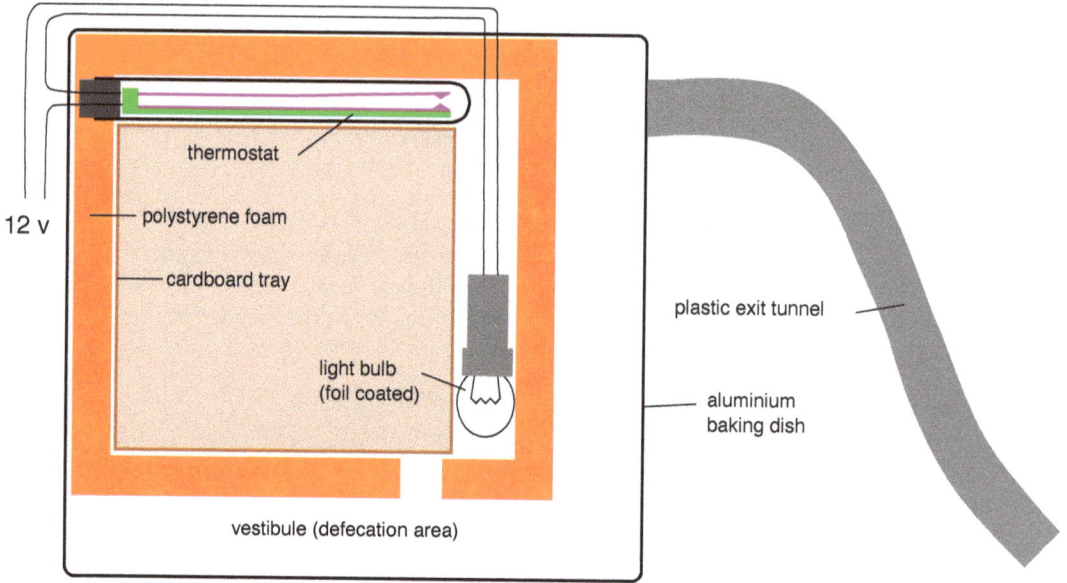

thermostat

12 v

polystyrene foam

cardboard tray

light bulb
(foil coated)

plastic exit tunnel

aluminium
baking dish

vestibule (defecation area)

Figure 2.2 More advanced light-bulb hive. Schematic view from above.

Figure 2.3 Small colony of *B. terrestris* in a light bulb hive. Initially, pollen was placed in a plastic cap on the cardboard tray, and the tunnel was connected to a box with honey-water. When more workers emerged, it was connected to a window for free foraging, so they were self-sufficient for pollen and nectar.

Small (5W) light bulbs gave a practical amount of heat, but it was unevenly distributed. A more even heat source was needed – something like an electric blanket. But electric blankets ran on 240 or 110 volts and were too big. So I tore up an old electric blanket, extracted the heating element, cut off a shorter length and powered it with twelve volts, which was safer (and legal) for amateur tinkering. Details of the power, voltage and cable specifications are given in Appendix 1. I did not want a fabric blanket because the bees would shred it, so I sandwiched the insulated cable between sheet metal and asbestos stove mat to form a flat warm-plate about twelve centimetres square. (In those days, we had thin disks of asbestos to use on stove tops to moderate the element heat and to catch spills.) These flat-plate heated hives still used fish tank thermostats, but as there was no need to protect them from water, and they were carrying a safe voltage, I replaced the glass tube covering with wire or plastic screen to speed up their reaction to temperature changes. I placed these sheet-metal warming plates on the floors of the foam polystyrene boxes I used as the first type of (unsuccessful) surface domiciles and also in the school-room experiment (see Figure 1.3). I made these flat-plate heated hives before officially starting my MSc thesis. They housed some of the *B. ruderatus* colonies I had discovered in the field, as described in Chapter 1.

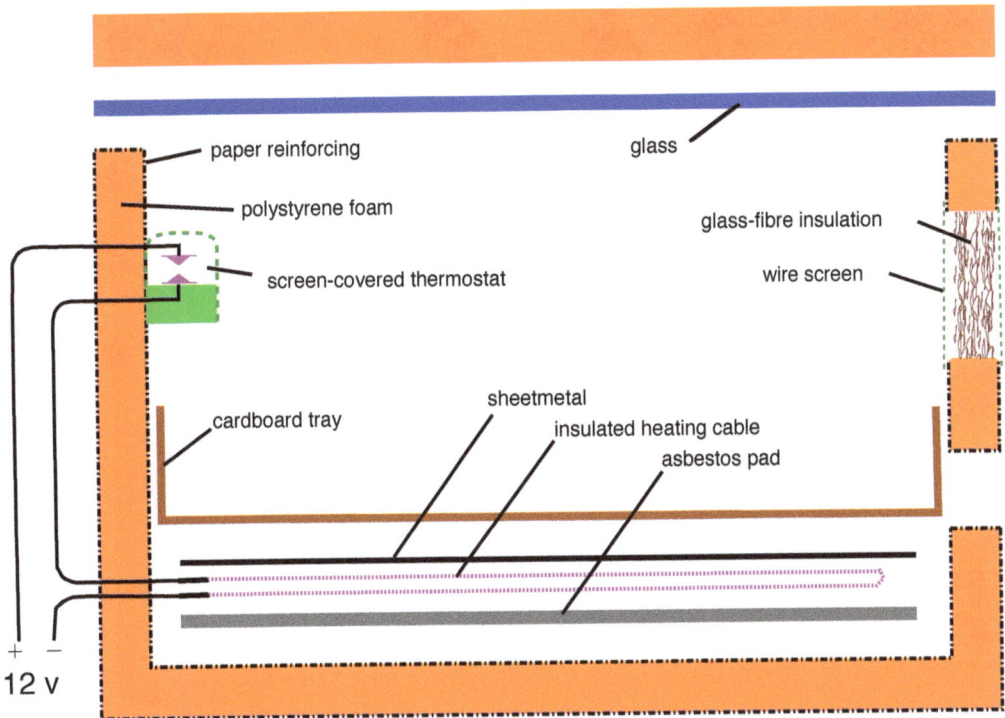

Figure 2.4 Flat plate heated observation hive. Vertical section expanded view. Internal dimensions: 160 x 160 x 100 mm (6.3 x 6.3 x 4 in).

Figure 2.5 Colony of *B. ruderatus*, excavated from the field, in a flat plate observation hive. Note that the bees have applied wax over the apertures of the thermostat screen.

The shape of bumblebee comb

Bumblebee colonies expand upwards and outwards from the first brood clump. Each cluster of cocoons becomes the platform on which several new brood clumps are formed, so the comb expands upwards and outwards. The top one or two layers are live brood, and the structure underneath consists of empty cocoons in which honey is stored. As the comb expands outwards, it is supported by the nest material, but if exposed in a flat-floored hive, the brood clumps topple outwards and become widely spaced. This is what I found in my early box-shaped hives.

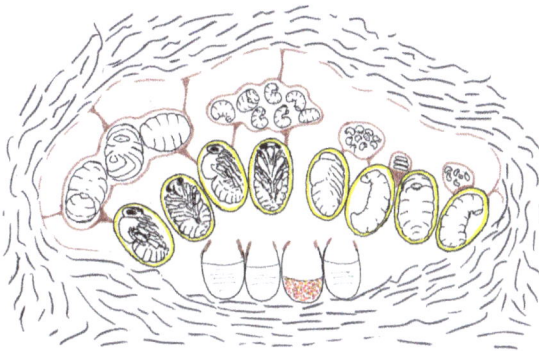

Figure 2.6 Drawing of a vertical section **through a bumblebee nest in natural nest material.**

Figure 2.7 Top view of a *B. terrestris* colony **in a cone-shaped observation hive. Figure 2.9 shows wider view.**

To accommodate naturally-expanding brood combs, I ideally wanted a bowl-shaped hive. But whereas a colony that has a slow growth rate and a short growth duration may end up the size and shape of a tennis ball or smaller, a rapidly growing, long-lasting colony may be more the shape of a wide kitchen bowl. The answer was to use a cone shape. There was no perfect angle: It would depend on the rate of sideways expansion as the colony grew, so I chose what 'looked right.' As a colony's expansion slows, the upper layers become narrower again, so I decided to have a vertical wall above a certain point.

Conical metal hive[3]

For housing the colonies I was expecting to get from the domiciles, I wanted hives that didn't contain bulky thermostats and were more suited to the shape formed by natural colonies in rounded earthen cavities. The design I developed is shown in Figure 2.8. The cone shape was intended to support the natural expansion of the brood comb (see 'shape' box), and the metal structure maximised the spread of warmth from the heating cable underneath. The cable is shown diagrammatically but was arranged either as a spiral or a pair of circular zig-zag loops around the cone. We used galvanised steel, but aluminium may be easier to cut and shape. When the first ones were made in 1975, perforated sheet metal was readily available, but when my graduate students needed more in the 1980s, the workshop technician chose to drill multiple 3 mm holes. I still use these hives, with upgraded electronics, nearly fifty years later (Figure 2.9 is a photograph taken in 2021).

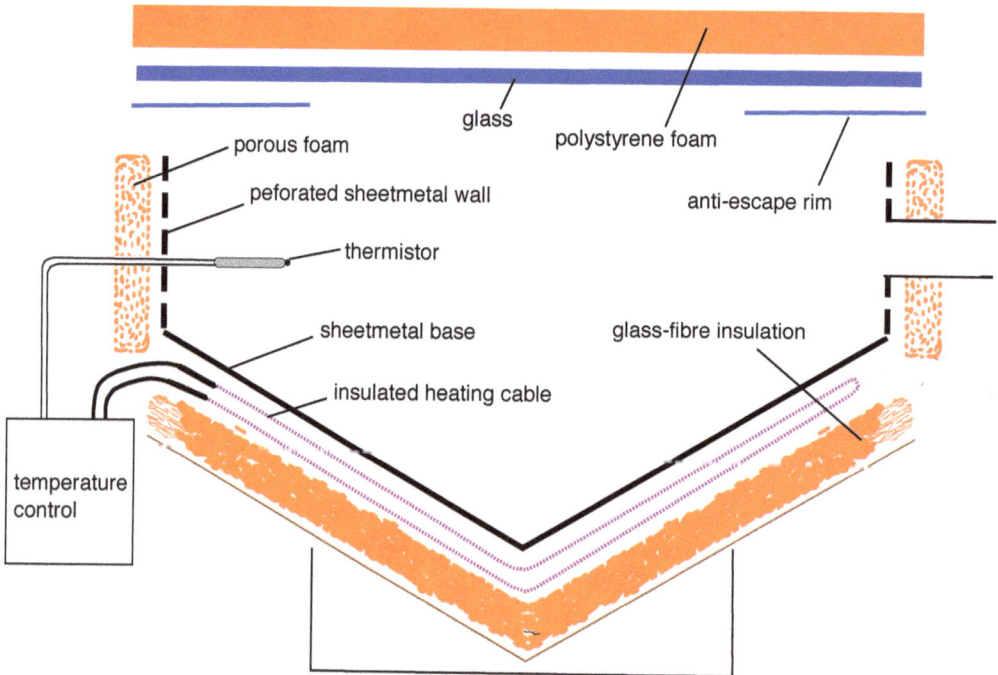

Figure 2.8 Conical metal observation hive. Vertical section, expanded view. The heating cable is represented schematically. It was arranged either as a spiral or concentric circles around the cone. The version built for my PhD studies in Toronto had a metal stud thermistor bolted to the metal base. Internal diameter 200 mm (8 in).

Figure 2.9 *B. terrestris* colony in a conical observation hive.
Bees exit to the wooden vestibule and forage via a wire
mesh tunnel (upper left). This later version has thicker foam
insulation covered with black cloth to exclude light.

Ventilation

Heating and ventilation tend to conflict. Your house cools down when you open the windows for fresh air. Most previous bumblebee hive designs typically used mesh-covered holes for ventilation. But if I was heating my hives, I didn't want cold draughts, and I had found that bumblebees in observation hives with mesh-covered ventilation apertures tended to cover the mesh with a layer of wax, presumably in response to light or coolness. I wanted a more uniform, porous airflow, such as a natural colony would have through the nest material and surrounding soil. My flat-plate heated hives had two 5 cm (2 in) square holes covered with wire screen on both sides of the 10 mm (3/8 in) wall, with fibreglass insulation sandwiched between (Figure 2.4). This seemed fine, although the workers still blocked some of the mesh apertures with wax. The conical hives achieved even more dispersed ventilation via numerous small holes. I experimented on vapour diffusion through various surfaces by measuring the evaporation loss from jars of water with multiple lid coverings. Taking into account the total surface area, the perforated metal rim provided similar vapour exchange to a nest covering of jute felt, whereas the fibreglass-padded large vents on the flat-plate hives gave 25%, and a pair of typical 25 mm holes covered with plain screen gave only 5%, of the vapour exchange of the more natural nest envelope. Bumblebees are well known to fan their wings to increase nest ventilation, most obviously when the hive overheats, but also in response to excessive humidity or carbon dioxide. But I felt fanning was a response to suboptimal conditions and did not wish the bees to rely on it.

Sanitation

Bumblebees poo a lot, mainly at the edge of the nest cavity and in the exit tunnel. In natural earth, the liquid soaks away, but on solid surfaces it quickly becomes a yellow swamp that gets smelly and becomes food for tiny fly larvae. Figure 2.10 shows faecal accumulation in the corner of a wooden box containing a large colony of *B. terrestris*. Two workers

Figure 2.10 Workers fanning near wet faeces.

are standing in place fanning, presumably to disperse the humidity. There is a newspaper fragment in the image – initially, when installing this large field-excavated colony, I had placed layers of newspaper on the floor. The workers subsequently shredded it and covered the comb with it, presumably for insulation.

My long-term wish has been to design a bumblebee latrine – some device they would be persuaded to use for the bulk of their defecation that would direct it into a reservoir where they could not walk in it. A BSc Honours student, Beth Williams, did controlled experiments that confirmed the preference for a rough substrate, internal edges and corners in rectangular spaces and cooler locations as defecation sites. The circular shape of the conical hives had the added advantage of no corners, but as an added precaution, I lined the conical floors with plaster of Paris-impregnated cloth, which gave a rough foot-hold as well as being moisture-absorbent.

The long-standing solution has been to insert a vestibule between the colony and its exit tunnel, and Sladen advised it to be lined with earth to be changed weekly. Others, like Free and Butler, suggested corrugated cardboard or paper.[4] These days, there are proprietary absorbent cat litter granules available. For my general colonies, I usually use a litter-lined plain wood or plastic box as a vestibule, but for the thesis research, I used 500 ml (one pint) round food pots with the floors replaced with a cone of plastic mesh. These units were designed to intercept larvae that the workers sometimes ejected from the colony. But that's a long story better told elsewhere.

Even with a vestibule in which most faeces are discharged, some defecation occurs in the tunnel beyond, so I also made these from mesh. The resulting mesh tube is quite flexible, provided the mesh is a simple weave (like many shade cloths and metal fly screen) and not welded at the intersections. You need to anticipate the ejection of yellow 'splots' from these tunnels and appropriately protect nearby walls or anything you don't want to be dirtied.

Access to the outdoors

With several colonies in one room, the obvious way to duct them to the outdoors is through a series of holes in the wall. But building managers are not keen on this, it can be difficult to make the exit surroundings distinctive enough for the workers to find their way to the correct hole, and curious passers-by may interfere (this was quite close to student residences). For the first season, I had ten colonies on a long bench; each ducted into a common cheesecloth flight duct (about 30 x 40 cm and 5 m long) attached to the pane-less window. This more or less worked, but despite the distinct decorations identifying each junction into the flight duct, I sometimes found marked workers in the wrong hive. It seemed that the cheesecloth duct was too confining for them to make accurate orientation flights.

In the second season I had a dedicated room, so I felt less need to confine the workers within a duct, and I let them fly throughout the room and out the window. I removed a whole upper window frame and placed a decorated

Figure 2.11 Aluminium screen exit tunnel. Screen tunnels remain much drier than those made from plastic tube. Tunnels can be two or more metres long and can be made by rolling metal or plastic mesh around broomstick and securing by binding with string or fine wire. Faeces often stick to the mesh but dry quickly.

cloth awning over the outside to give rain protection and to help the workers recognise their window. I gave the colonies 2.5m (8ft) tunnels with the exit points scattered over a wide bench area and decorated with various objects. I needed to be careful if working there at night because workers would sometimes go out of their tunnels and fly to the electric light.

Monitoring the colonies

I intended this book to be mainly about ways of working with bumblebees. Most of the resulting data and conclusions about bumblebee biology are best told elsewhere. Still, I will be giving some results to show the sorts of information I was able to gather.

Foraging activity

Electronic technology has long superseded my old tricks. I first tried to monitor the forager movements of a small *B. hortorum* colony (from the school room experiment page 18) by placing a bee-weight-activated electric switch in

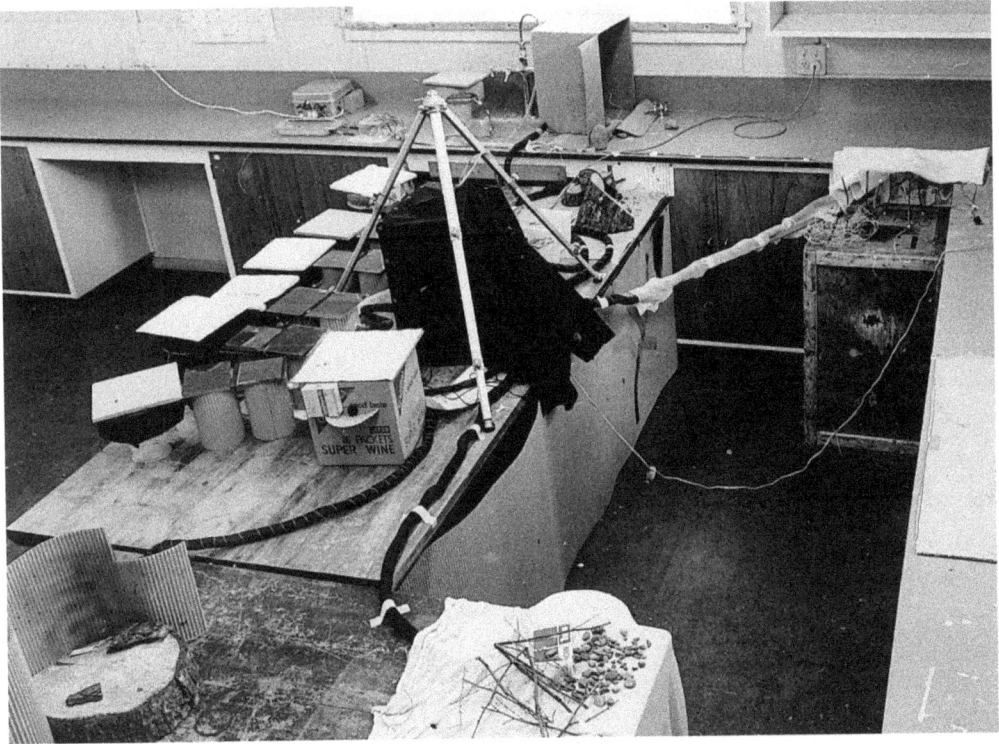

Figure 2.12 Laboratory layout with six observation hives. Plastic mesh exit tunnels, 2.5 m (8 ft) long, terminated on the bench by distinctive 'decorations'. The foragers flew loose in the room and out a high window.

the tunnel. It was made from a postage-stamp-sized piece of balsa wood and copper. When tipped by a passing worker, it switched an electromagnet (hand wound on an iron nail) that moved a lever to scratch a mark on a home-made smoked drum. If the drum went slow enough to give a day's trace, the scratches were too close together to separate, even though the scratch point was a prickle from a thistle – the finest point I could find. It was a fun challenge to make during my holidays before starting the MSc proper, but – like so many endeavours – not successful the first time.

My aim for the thesis work was to estimate daily food intake from measurements of foraging activity. During my first research season, the foragers passed through glass tubes, breaking a light beam and triggering electromechanical counters. The sensing system was fine, but electronically there was interference, and the counters sometimes clicked merrily in the absence of bees. The following season – the main year for my thesis data – I arranged the long exit tunnels to converge on six parallel channels with light sensors in the floor and a red lamp above. When a bee shaded any one sensor, a Super-8 movie camera shot a single frame illuminated by an electronic flash. Sample frames are shown in Figure 2.13. Note my old-school method of providing a time and date stamp: a wristwatch and a hand-written date. The idea was to measure pollen load sizes from the images, and by removing and weighing a subsample of pollen loads, calculate an image-size:weight relationship from which the daily pollen intake could be estimated. I got as far as examining about two thousand frames from one film at 40x magnification under a microscope to measure pollen load sizes (they did not all have pollen loads – half the images were bees going out, and some returnees had no pollen), but found it too time-consuming to process the other five films in this way. Instead, I spooled them through a frame scanner and merely recorded the presence or absence of pollen loads, obtained data on nineteen colony days and was able to show a relationship between the number of pollen loads entering a colony and the quantity of larvae present.

Only later, when doing the red clover pollination trial, did I develop the idea of a 'trapping vestibule' as a way to measure the standing crop of foragers in the field from a colony. It would not have told me the daily pollen intake but it became standard fare for measuring foraging rates for my pollination research (Chapter 3).

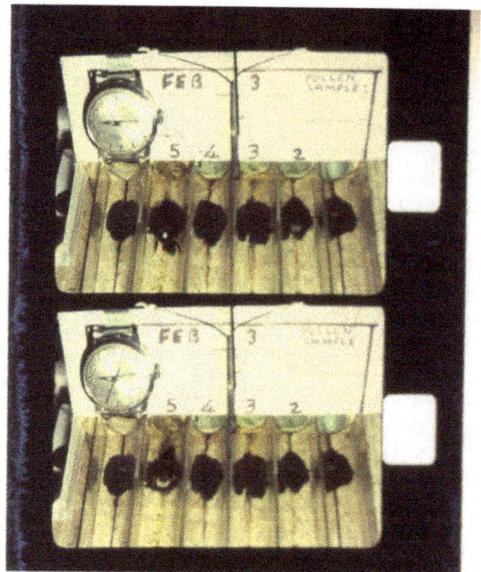

Figure 2.13 Two still frames of foragers from Super-8 film. A video camera was set to take single images of *B. ruderatus* foragers as they passed over a light sensor. The upper frame is a marked queen leaving, and the lower one is a returning worker.

Individual food intake

I did not know if anyone had measured how much pollen was needed to rear a bumblebee of a particular size, and there was also the question of whether queen larvae were fed anything different, like 'royal jelly.' I thought if there was an alternative food for queens, they would consume proportionately less pollen than a worker

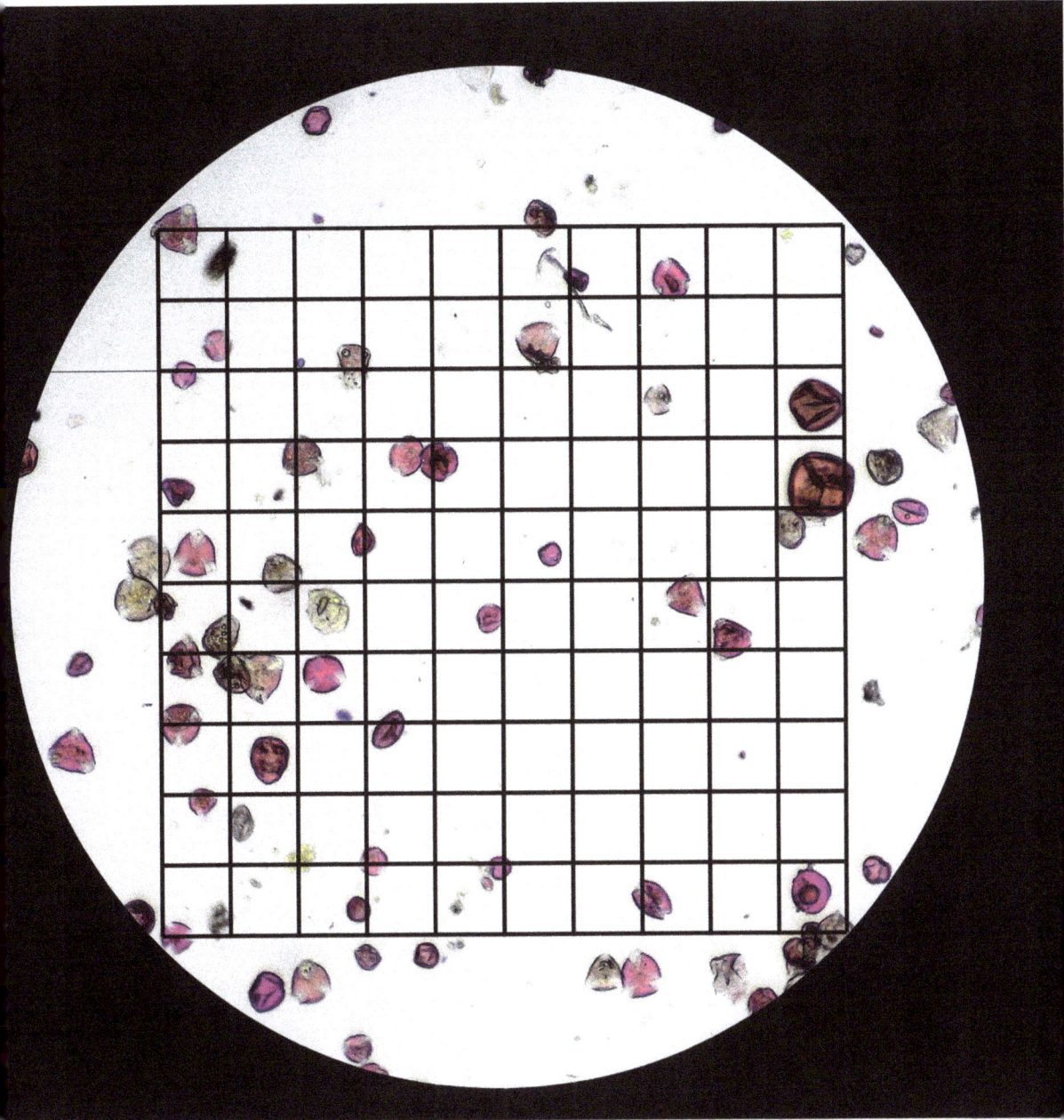

Figure 2.14 Pollen grains under microscope. After dissolving the silk, what remained were the empty walls of all the pollen grains eaten by that larva during its growth. With suitable dilution they could be counted in fixed-volume cell, including in most cases identification to flower species.

for their size. It is not easy to measure the pollen consumption of an individual larva unless you feed it 'by hand' in the absence of nurse bees. I used another method: faecal analysis. Pollen is a marvellously rich food; the only indigestible waste is the outer coat of each pollen grain. Because the waste is so minimal, bee larvae do not defecate during their growth – all the pollen grain coats are compacted in their hind gut, taking little space like a bin of compacted bread bags. When the larva has finished its growth and is spinning its silken cocoon, it finally eliminates a series of dark soft faeces ('meconium') which become embedded in the lower half of the cocoon. This distribution means the meconium is not disturbed when the bee makes the exit hole at the top. This probably has sanitary advantages, and the meconium does dry and harden, strengthening the comb structure and would be more difficult for a bee to chew through. The other marvelousness of pollen grains (for my purposes) was that the coats are very resistant to chemicals, so I could use corrosive alkali to dissolve away the silk to recover the pollen grains. (Pollen gains are so durable that they remain identifiable indefinitely in peat bogs and other sediments.)

So, given an empty known-diameter cocoon from which a known-sized bee emerged, I could retrieve all the empty pollen grain coats it ever ate by dissolving the cocoon structure in potassium hydroxide. After sufficient stirring and dilution, the pollen grain coats could be examined in a counting chamber under a microscope. After the silk-dissolving treatment, the grains resumed their original shape – indeed, they looked no different than fresh ones, and it was hard to believe all their internal nutrients had been digested. So by calculating dilution rates, the volume under the microscope grid, and by representative counts, I calculated the total number of pollen grains the larva had eaten.

But I needed to convert a result of, say, 14.5 million pollen grains into a weight as collected by foragers. Pollen grains varied in size between flower species, so I needed to work out the weight-per-million-grains for the common species separately. Fortunately, pollen grains have distinctive shapes and surface sculpturing, so it was possible to assign floral sources to most of those I saw under the microscope. From seeing what flowers in the area were being visited by *B. ruderatus* workers and sampling the pollen, I was able to match up about five forage species (predominantly red clover and goat's rue) with pollen loads on the bee's legs, which I removed, weighed, and also counted under the microscope. For example, one gram of red clover pollen comprised 16 million grains.

Putting it all together, I could show that worker larvae consumed 0.15 to 0.6 g (workers vary greatly in size) and that queens consumed 0.9 to 1.1 g of pollen. Using a range of cocoon sizes, I could show a direct linear relationship between adult bee size and the amount of pollen consumed. This suggested that queens obtained all their protein nutrition from pollen, the same as workers and males.

It was evident from observations that newly emerged adult queens also ate a lot of pollen. I could see them feeding on it, and it showed copiously in their faeces, but only for their first few days. So I removed some groups of male and queen cocoons from a colony, let the adults emerge in an incubator away from other bees, and measured the amount of pollen consumed (how much weight was lost from small pots of pollen). A typical queen consumed an additional 0.3 g of pollen over its first few days of adulthood.

'Productivity Index' as a measure of colony strength

What is the ideal measure of 'how big' a colony grows? The strength for pollination, at any one time, should be the number of workers actively foraging on the target crop. I will get to this in the next chapter. Later, when there was a commercial bumblebee industry, colony strength was often quoted as the number of bees in the hive, although they were seldom directly counted. I wanted a measure that would reflect the foraging work done over the life of a colony. I had counts of the empty cocoons, but the problem was that male and worker cocoons were similar sized. Males do not forage for the parent colony and are irrelevant for many pollination applications. So I thought: instead of viewing an empty cocoon as a producer of an adult, which may or may not have done any foraging work, I should look at the cocoon as a measure of the foraging work already done to rear it.

Having measured the relative amounts of pollen required to rear a worker and a queen of *B. ruderatus* and found that a queen requires 3.3 times as much pollen as an average worker, I decided to count all the small (worker and male) cocoons and give them a value of 1.0 each, and all the queen cocoons and give them a value of 3.3 each. Summing these weighted values would provide a figure proportionate to the total pollen consumed, and hence the foraging effort, of that colony. I called this figure the 'Productivity Index.' It was intended as a relative figure to compare colonies, and not an actual measure of the total pollen consumed, as an additional amount passes through the adults' guts and appears in their faeces without being fed to larvae.

Overall colony growth

My main interest was what was happening within the nest regarding brood development and the sex and caste of emergent adults. I mapped the brood clumps daily by drawing their shapes on transparent sheets placed on the glass lids of the observation hives and took daily photographs. Each cluster of egg cells was given a number, and its development was followed with notes on when cocoons were formed, when adults emerged, and the sex/caste of those adults. When the colony had died out, I could identify each cluster of empty cocoons by referring back to the photographs, counting them, and thus account for every bee that emerged from each colony, with an estimated date of emergence. To retrospectively determine the size of every bee, I measured each cocoon. The technology has been superseded now, but then we had 'overhead projectors' with a flat glass bed intended for the placement of transparent sheets to be projected onto a large screen. I placed individual cocoons on the bed and projected their shadow onto a screen. Some were distorted, but with a set of standard circles calibrated to equate to cocoon diameters, I could get a reasonable estimate of the size of most cocoons. I had separately captured and measured some individual workers that emerged from known cocoons, and I was able to show, unsurprisingly, that bee size was directly correlated with cocoon diameter.

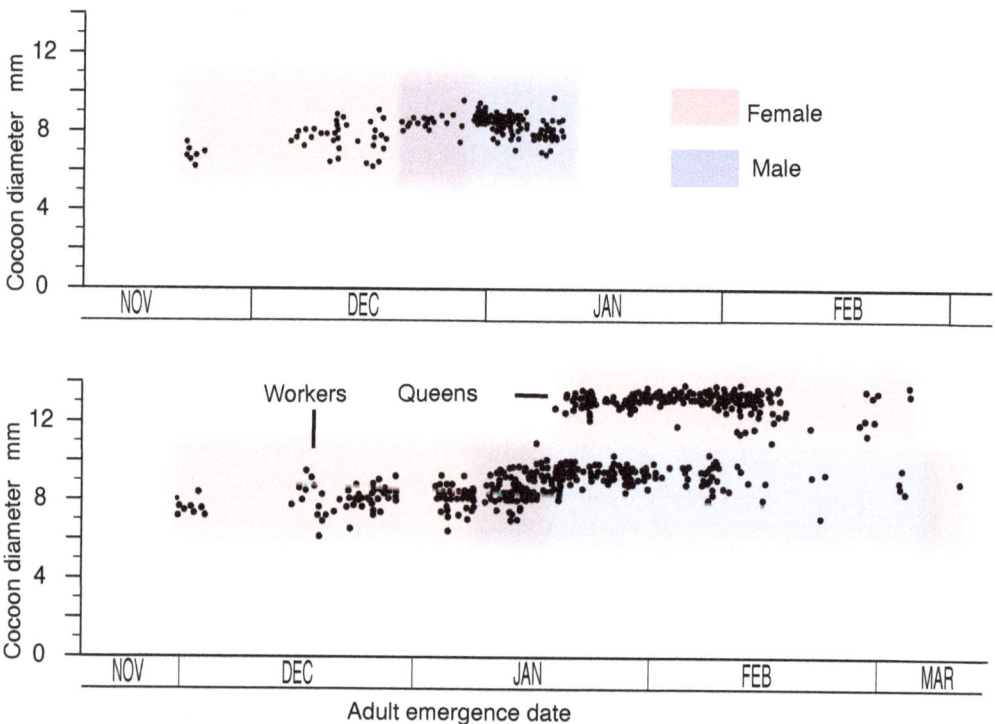

Figure 2.15 Graphs showing the adult production for two *B. ruderatus* **colonies.** Each dot represents one adult identified and sized from the empty cocoon at the end of the season. Colony 1 (upper) had the shortest period before male production and produced 237 adults but no queens. Colony 5 (lower) produced 662 individuals, including 228 queens.

In Figure 2.15, each dot represents an adult bumblebee, its sex, date of emergence, and size, all collated from the regular brood mapping and measurement of the empty cocoons after the colony had declined. The smallest and the largest colonies are shown. The small colony produced no queens and was the earliest to begin male production. I knew at what time adults emerged from each brood clump, but only the presence or absence of newly emerged adults of each sex, so I made the approximation of equal numbers of each when both were emerging. The total number of workers thus estimated to have been produced by the colonies was directly proportional to the duration of the worker production phase ($R^2 = 0.99$, linear). This is hardly a surprise; in turn, the total productivity was also proportional to the duration of the worker production. In this case, an exponential relationship gave a better fit ($R^2 = 0.88$). See Figure 2.16.

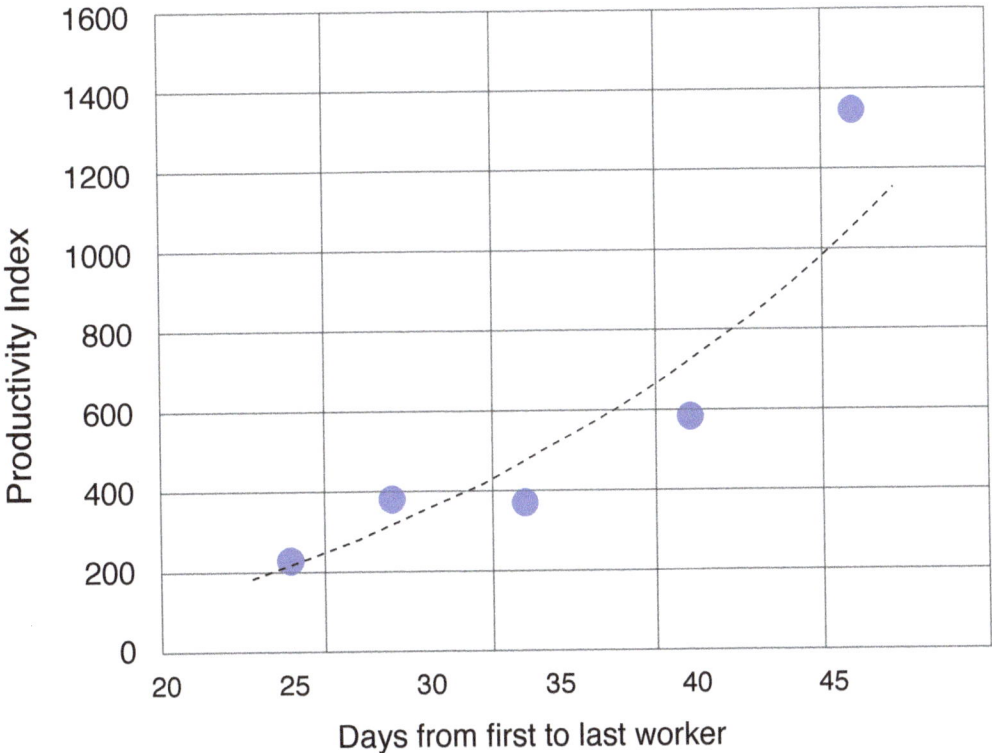

Figure 2.16. **Graph of productivity vs duration of worker production.**

My interpretation of much of the literature I had seen at the time (the mid-1970s) was that bumblebee colonies met their demise either by the onset of winter or by the change to queen production due to surplus food economics as the worker population expanded.[5] However, my observations of the B. *ruderatus* colonies suggested that the transition from worker to male production was the critical factor. The work by Peter-Frank Röseler in Germany also indicated that the timing of male production was important, and he pointed to dominance interactions between the queen and workers.[6] Another tantalising idea came from Bill Hamilton's work on genetics and kin selection which predicted antagonism between the queen and her workers when males were produced.[7] I looked at some of this in my subsequent thesis.

Bumblebees in Toronto

Another significant 'connection' arose through my writing to Chris Plowright at the University of Toronto (U of T). Chris was one of the world gurus of bumblebee research at the time (mid 1970s), and he was highly successful at rearing colonies of bumblebees.[8] He reported keeping his colonies at 30°C (86°F), and I wanted to know what heating system he used. It turned out he just placed them in an incubator cabinet. But anyway, he invited me for PhD research in his lab. This was momentous for me: I had an erratic undergraduate record, and even students more successful than I were not considered likely to win the necessary scholarships to study overseas. But Chris promised some creative financial arrangements that would make it possible. And there were two of us by then – I'd married Rachel the previous year. One way and another, with practical support from others, we kept ourselves going for our nearly four years in Canada. We produced our first child there.

Chris was a tall man who never walked slowly. Some students found him intimidating, and others saw him charming, but although I have a much more docile personality, there was some kindred spirit between us. Forty-five years later, we occasionally have animated phone discussions, and I would have visited him in Ottawa in 2020 had the Covid pandemic not scuppered my travel plans.

One of Chris's lab's delights was his importation of bumblebee species from around the world. Before I left New Zealand, I had air-freighted him a few queens of each of the four species here. It was cute to see little colonies founded by queens Rachel and I had captured at home a couple of months earlier, including the elusive *B. subterraneus*, which features in Dave Goulson's book *A Sting in the Tail*, where he gives a comprehensive account of attempts to re-introduce the species to Britain after it had become extinct there.[9] (More on this species in Chapter 8). There were strict conditions, including that the bees and their progeny were kept confined indoors and never allowed outside the lab.

More observation hives

Chris was keen to use the electrically heated conical observation hives, and I helped assemble the electronics (revised, with a more sophisticated proportional controller). Chris's group needed even more observation hives, but the electric ones were expensive. His standard hive had been a plywood box with a cardboard conical insert with cotton insulation between the insert and the walls of the box, and these continued to be used. But we liked the idea of being able to cast an object, circular with a conical depression, from plaster of Paris or concrete, and we visited U of T's Engineering Department to learn more about porous, lightweight concrete. Pumice was not readily available so far from the volcanic regions, but we learned that expanded perlite would be a good substitute and was even lighter. We also learned that gypsum (the product of setting plaster of Paris) could be made more water-resistant by mixing it with some Portland cement. Since then, I have done a lot with the plaster-cement mixture: it sets even faster than plaster alone and makes a hard product more like concrete. So we decided to cast observation hives from a mixture of perlite, plaster and cement. Mixed to a stiff paste with water, it set enough to handle in a few minutes. The moulds were made by the Zoology workshops, lathe-turned from laminated plywood, and varnished. We sometimes had up to a hundred colonies in perlite hives lined up on teaching lab benches during the students' summer vacation.

Figure 2.17 Colony of *Bombus impatiens* in a cast Perlite-cement observation hive. **This was one of the dozens in a teaching lab during the summer vacation. Sugar-water was dispensed in tubular gravity feeders.**

Where it (almost) led

My thesis work at Toronto was about the inner life of the bumblebee colony – the interactions between the queen and workers – what I called 'reproductive dominance'. My ambition to become a bumblebee keeper was shifting more towards bumblebee biology - understanding what limited colony growth - as a prerequisite for managing bumblebees economically. I was forming ideas that blended the earlier ideas of John Free and others that colony maturity was an economic phenomenon (e.g. worker: larva ratios and food surplus) with newer ideas, especially from Peter-Frank Röseler, that it was more a social phenomenon where the workers caused stress to the queen as their numbers built up. It seemed that these could interact to create a 'socioeconomic' picture, and I was keen to measure a pollen budget to see how food surpluses affected worker physiology. I was accepted to a postdoctoral fellowship at New Zealand's Otago University to pursue this.

The 'bumblebee keeper' picture was still alive, though, and although in Chris's lab, we were proud that one person could look after two-hundred incipient colonies, I thought there should be a way that a person could look after two thousand. That thread resumes in Chapter 4.

Chapter 3

Bumblebees in the crop and orchard: the pollination numbers game

Bumblebees in the crop and orchard: the pollination numbers game

After defending my PhD thesis early in 1981, I returned from Toronto to New Zealand. Rachel had travelled a few months earlier, before our son turned two, and would have required an airfare. I was due to begin the postdoctoral fellowship at the University of Otago, but we enjoyed being reunited with our families spread around the Manawatu-Whanganui area and the prospect of future extended family support. We procrastinated over the move to far-away Dunedin, and between labouring jobs with my brother Eric, I looked for other opportunities. Wanting to push for a scientific career in pollination with bumblebees, I took an overnight train to Auckland and managed to get meetings with a senior figure in the kiwifruit industry and with the director of the Entomology Division of DSIR[1] (the same organisation I'd had a summer internship with eight years earlier). I was hosted with good grace but was told the kiwifruit industry took its scientific lead from government research institutes. So I went back to building farm fences up the remote Waitotara valley.

One day, my father made the long drive up to the valley work site to leave a message that I should phone Brian Springett, then head of the Department of Botany and Zoology at Massey University, for 'information to my benefit'. Prof. Springett offered me a post-doc at a salary of $18,000. Otago's post-doc paid $10,000. With awkward apologies to Prof. Don MacGregor at Otago, who'd offered me the place there, I took the Massey position.

I thought bumblebees might be a lost cause, and I considered various lines of insect research, mainly concerning pasture pests. But I knew there had been ongoing studies on bumblebees for red clover pollination at Lincoln, near Christchurch in South Island (Entomology and Grasslands divisions of DSIR). And I heard from Rod Macfarlane, a long-time stalwart of the subject, that some red clover growers were buying hives of bumblebees, and there was talk of some kiwifruit growers being offered them for $100 each. This astonished me and suggested the subject was still very much alive and I should get into it.

The pollination numbers game

I was intrigued that anyone would pay $100 for a hive of bumblebees with no more than a few hundred bees, when hives of honeybees with tens of thousands of bees, were hired for less. I had yet to see any comprehensive cost-benefit analyses of producing bumblebees for pollination and was sceptical of some stocking rate calculations.[2] I was a junior academic on a one-year contract, and foresaw that after the university fellowship I might become a bumblebee keeper. But first, I needed to get a grip on the economics.

Farming is a business: crops are expected to be profitable. Yields vary for a variety of reasons. For a pollination-dependent crop, under a particular environment, it is fair to assume there is a positive correlation between the number of pollinators and the crop yield. Still, the trend may not be linear and must be asymptotic. It must flatten off as the seed number or fruit size has reached its biological maximum, i.e. pollination reaches saturation. Where the farmer has control over the number of pollinators, it should be most profitable to aim for a yield near the top of the curve. Everything has a cost in time and resources, so the question for bumblebee pollination is, 'how many bumblebee colonies would be needed to produce a commercially acceptable yield of this crop?' Then the second question is: 'What would that number of bumblebee colonies cost?' The bumblebee cost could be for environmental enhancement to increase the wild bumblebee population, or be the purchase price for commercially produced bumblebees.

The above questions are simple, but the answers are complicated and rarely settled at a scientific level. Researchers can easily measure the cost-benefit pattern of something like fertiliser by setting out a series of identical plots and applying a range of fertiliser concentrations, thus deriving a dose-response curve. But bees fly loose, so vast areas of the countryside would be needed to set up a controlled experiment with a range of hive densities. Cages have sometimes been used, but usually only to see the effect of saturating a patch of the crop with very high densities of bees.[3] Another approach has been to determine how much seed or fruit one bee would set based on how many flowers it visits and the effectiveness of each visit.[4] The performance of one bee is then extrapolated to the pollination performance of a whole colony. The obvious question of how many bees are foraging from one colony seldom seems to have been directly measured.

In this chapter, I describe my work with red clover and kiwifruit pollination. The red clover work has not hitherto been published, and I have gone into detail as a case study of how pollination stocking rates may be estimated. The kiwifruit work has been partially published, with a fuller account in an industry magazine.[5]

Red clover

Pollination of red clover (*Trifolium pratense*) was the purpose for bumblebees being introduced to New Zealand from Britain in 1888 and 1905.[6] Darwin had earlier shown the need for cross-pollination in this plant and concluded that only bees with long tongues would be able to reach the nectar at the bottom of the long tubular florets.[7] A survey around the middle of the twentieth century confirmed that four species were established: three similar-looking long-tongued species, *B. hortorum*, *B. ruderatus*, and *B. subterraneus*; and the short-tongued *B. terrestris*.[8]

Figure 3.1 (top). A red clover seed production crop in the Wairarapa region on the North Island.

Figure 3.2 (above). Vertical section through a red clover flower

Figure 3.3 Single floret. **Upper: intact. Lower has had lower petals removed to show stamens and stigma.**

The flower depth–tongue length story has appealing logic but overlooks the fact that short-tongued bumblebees and honeybees can readily access the pollen without reaching the nectar, and can and do often collect pollen by itself, thus pollinating the flower. Nevertheless, nineteenth-century farmers growing red clover for seed were dissatisfied with their yields and reported significant increases after the bumblebees were introduced. But by the mid-20th century, it was clear that seed yields were well below their potential. The newly-developed, more vigorous tetraploid cultivars were considered to be even more dependent on long-tongued bees due to their large flowers. The farmers were harvesting as little as 50 to 100 kg of seed per hectare, whereas yields as high as 1000 kg/ha (890 lb/ac) were possible with complete pollination. The shortfall tended to be worse when red clover was grown in large fields – there just wouldn't have been enough long-tongued bumblebee colonies within flying distance.[9]

Work by G.A. Hobbs in the 1960s in Alberta, Canada, involved placing domiciles in the field.[10] He intended to procure colonies in the Rockies foothills, where long-tongued species were common, and transfer the colonies to the crops at flowering time. This was also intended in New Zealand, but with the more compact landscape, domiciles were often placed near the crop. If the primary limitation on the wild population near the crop was a lack of nest sites, then there was a chance that merely placing domiciles nearby would boost the population. Despite the additional nest sites, I doubted that typical cropping landscapes would support a sufficiently high density of bumblebees for maximum crop pollination due to the relative dearth of floral resources earlier in colony development.[11] As with procuring colonies for my MSc thesis work, I thought it more practical to place domiciles where *B. ruderatus* was naturally very common, and to move the colonies to the crop at the time they were needed for pollination.

I mentioned in the previous chapter how I had made contact with the seed company Wrightson NMA, and they had given me a small donation towards my thesis research. I renewed the contact and got more substantial funding to investigate how many bumblebee colonies would be needed to pollinate an area of red clover seed crop.

Design of the trial

I obtained access to four red clover seed crops, which ranged in area from one-fifteenth to about one-and-a-half hectares (0.16 to 3.6 acres).[12] I could not directly control the density of bumblebees on the crops – there were already wild bumblebees present, and I could not control how widely my bees dispersed. So, I aimed to at least create a wide range of bumblebee densities by adding various numbers of hives containing *B. ruderatus* colonies at flowering time. One crop received no hives, and the highest density was approximately 300 hives per hectare (19 hives on a crop of 0.065 ha (125 hives per acre)).

I aimed to produce a four-point 'curve' for seed yield versus *B. ruderatus* forager density, ranging from very few to saturation. The critical figure was the number of bumblebees working the crop – the number of colonies introduced was incidental. Independently, I found a way to measure the number of red clover foragers working from a sample of hives. From these two parameters, the actual stocking rate could be derived from the relationship:

(hives needed per hectare) = (foragers needed per hectare) / (foragers per hive)

As an open-field trial I also needed to account for other types of bees: short-tongued bumblebees (*B. terrestris*), and honeybees, which were also present in varying numbers.

Obtaining the bumblebees

I was planning to procure around two hundred colonies of *B. ruderatus* for the trial, and constructed 350 of the flat-slab concrete domiciles (Figure 1.9). I installed 250 of them on the home farm and 100 on my brother's farm. Red clover seed crops were usually harvested in late summer, and I thought there was a chance that many of the *B. ruderatus* colonies naturally initiated on the home farm might be too old and in decline by the time they were needed for pollination. (The colony I observed from there as a seventeen-year-old produced queens in January: Figure 1.1). Hence the use of my brother's farm further inland and at a higher altitude (approximately 350m (1100 ft)), assuming the bumblebee season would be later there. My records show I spent most of September constructing the domicile parts and most of October installing them in the ground. Figure 3.4 shows a colony in one of the domiciles. As stated in Chapter 1, around 200 (58%) were occupied by *B. ruderatus* queens that produced workers.

Whereas the domiciles needed to be attractive to searching queens, and give a secure environment for a growing colony, the hives used to transport and house the colonies at the crop needed to be simple and lightweight. I was concerned about ventilation and faeces dispersal. Bumblebees produce copious wet faeces (see Figure 2.10), and I thought a wooden box would get too wet inside, so I made 20 cm cubes from aluminium fly-screen, supported on all edges by strips of wood about 10 mm (nearly half an inch) square in section. In retrospect, I was too sensitive to those requirements. In later years I found that simple cardboard cartons with suitable rain covers were adequate crop hives.

Figure 3.4 Concrete slab domicile with *B. ruderatus* colony. **Note the black mesh 'cradle' to ease removal. The nest material has been opened. A cell of eggs is visible.**

57

The colonies, supported by the mesh 'cradle' (Figure 3.4), were relatively easily transferred to the mesh crop hives. However, it was still demanding—driving over the hill farm tracks (Figure 1.2) in the dark, followed by the two-hour drive to Palmerston North. It was a recipe for falling asleep at the wheel, but fortunately, the Land Rover's noisy, harsh ride was far from soporific.

Figure 3.5 Portable crop hive made of aluminium screen. **Twenty cm (8 in) cube.**

Counting bees on the crops

On each of my four crops, I marked out ten plots of 2 x 2 metres, which I judged to be a manageable size to scan for bees. Within each plot, the numbers of each type of bee (long-tongued workers, males, queens; short-tongued workers/males, queens; and honeybees) were counted several times over several days. Fortunately, males of the long-tongued species could be distinguished by their longer antennae. *B. terrestris* males look more like workers.

Figure 3.6 *B. ruderatus* worker on red clover.

As bees work flower-to-flower, the most relevant measure of pollinating activity should be bee density in terms of the number working per a fixed *number of flowers*. The seed grower measures yield per unit *area*, but the two measures can be reconciled if the number of flowers per square metre is counted.

Counting the foragers from the hives: The trapping vestibule

I needed to relate the number of bumblebees I counted foraging on the crop to the number that came from a typical colony. To my knowledge, there are few published figures of direct counts of a bumblebee colony's foraging force to this day. Most reports of bumblebee foraging activity have been based on counting nest entrance traffic, assuming a proportion of the total workers in the colony will be foragers, or retrospectively estimating it from empty cocoon counts. These methods are time-consuming, retrospective, and/or require assumptions that are difficult to verify. I wanted a direct count of the foragers working from a colony. Having used one-way entrance/exit 'valves' using projecting smooth glass tubes in my early thesis work, I figured there must be a way to intercept returning foragers for examination and counting. My reasoning was this: at any moment, a certain number of workers are foraging outside the hive, and they all come back within a finite time, so the trick would be to stop any more going out, intercept the returning foragers, and count them when they had all returned. My previous observations suggested twenty minutes was a typical foraging duration, and I chose an hour as the time within which I assumed they would all be home again.[13]

This was done in an entrance vestibule – a two-chamber wooden box strapped to the hive's entrance before opening it for the first time on the crop. Workers leaving the colony passed through the two-chamber vestibule via a partition with one-way 'exit' and 'return' paths. The directionality was achieved by smooth glass or plastic tubes about 16 mm in diameter and 50 mm long (5/8 x 2 in), which either pointed in or out. Bumblebees went out through one path and returned through the other. When I wanted to measure the foraging population, I plugged the main hive so no more foragers could get out into the vestibule and plugged the 'exit' tube so returning foragers could not leave a second time.

I was being supercautious with this early design, not wanting to change anything about the foragers' usual paths. I thought that if anything varied about their return path, they might fly away again. With later designs, I was less cautious and simplified the design to a single chamber, so returning foragers would discover a blocked entrance and would need to enter via the adjacent one with the one-way tube.

Figure 3.7 Diagram of trapping vestibule. **Left side shows normal free-foraging; right side shows two blocked apertures, thus trapping all returning foragers. Since foraging duration rarely exceeds an hour, after that time the vestibule should contain all the foragers working at the time it was set.**

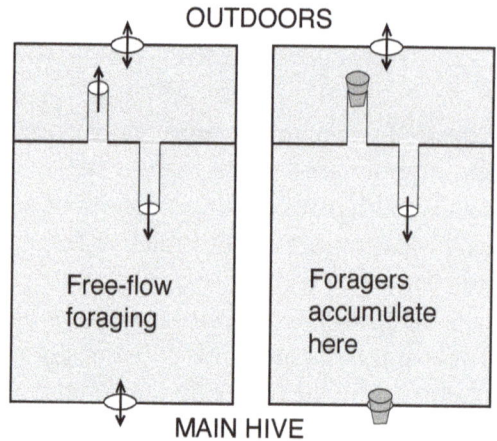

OUTDOORS

Free-flow foraging

Foragers accumulate here

MAIN HIVE

Figure 3.8 Trapping vestibule attached to a crop hive. **The wire screen hive has been wrapped in thick jute cloth to exclude light so the bees can find the exit more easily.**

It only took me a few seconds to set a vestibule for trapping, so I could do them all in succession. When an hour had elapsed after setting the first one, I could return around the vestibules, visually counting the trapped foragers and recording the presence and colour of corbicular pollen loads. After counting, the plugs were removed, and foraging resumed. This disrupted the colony's activity for a while, but I only did it at intervals of several days. Two crops had nineteen trapping vestibules each. The size of the foraging population was measured on two occasions, at a similar time of day to which the bees were counted on the crop (late morning, early afternoon).

Measuring seed set

Ten flowers from each plot were tagged and retrieved later for seed counting. The red clover flower head is capable of setting one seed per floret. Florets are recognisable on withered heads, and the presence or absence of a seed in each one can easily be felt, and the husks can be abraded to release the seeds. Four plots each on two fields were hand-harvested and run through a small-scale threshing machine to assess the harvestable seed.

Figure 3.9 Red clover flower head withering, and seeds.

Results: Number of bumblebees needed

Although total bee numbers varied considerably between the fields, the field with the fewest bumblebees coincidentally had fewer flowers, so the actual density there (i.e. bees per thousand flowers) was not as low as expected. Thus, the pollination result in mean seeds per flower covered a relatively narrow range (61 – 84 seeds, or 52 – 63% of florets, per head). There was a positive dose-response result for long-tongued bumblebees, and a negative association with the numbers of short-tongued bumblebees and honeybees. The results are shown in Figure 3.10.

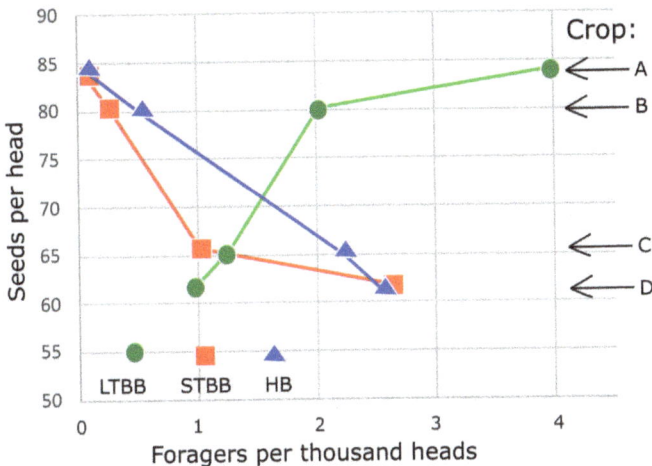

Figure 3.10 Graph of seed set in relation to bee densities. LTBB = long-tongued bumblebees, STBB short-tongued bumblebees, HB = honeybees. Beware: This gives the impression that STBB and HB *reduce* the seed set, but see the explanation in the text.

The negative trend in seed set with the density of short-tongued bees does not necessarily mean they were inhibiting pollination. You can see that crops A and B, with the most long-tongued bumblebees, had the least numbers of honeybees and short-tongued bumblebees, whereas the converse occurred with crop D. It is plausible that the long-tongued bumblebees were so effective at depleting the pollen and nectar, that flowers on those crops were less rewarding for the short-tongued species.

Crops C and D had moderate seed set even though long-tongued bumblebee numbers were low, so I assume the short-tongued bumblebees and honeybees, present in higher numbers, were doing some of the pollination. I did not try to fit a multiple regression to the data, but I did tinker with giving *B. terrestris* and honeybees partial values to see if the combination of long-tongued bumblebees plus de-valued short-tongued bumblebees and honeybees gave a tighter fit to the relation with seed set. Indeed the combination of (long-tongued bumblebees) + (short-tongued bumblebees /20) + (honeybees /10) gave the best fit of a range of adjustments I tried (Figure 3.11). I would not pay too much attention to the exact figures – honeybees may not provide twice the pollination of short-tongued bumblebees – the point is that short-tongued bees seem to have been making a positive contribution despite the negative interpretation that could be interpreted from Figure 3.10.

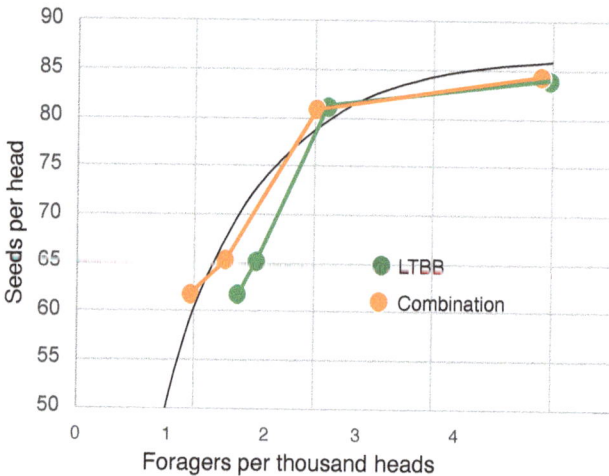

Figure 3.11 Graph of combined bee densities vs seed set. **Black line is an asymptotic curve with maximum = 88, green line is long-tongued bumblebees (LTBB) only, and brown line is LTBB + (STBB/20) + (HB/10).**

Although pollination seemed to plateau in crop A, the seed set per flower head was only 63% which is unlikely to be a biological maximum. However, previous studies suggest that tetraploid red clover seed set may fail to reach 100% even under optimal pollination due to other constraints.[14]

Results: Number of foragers from a bumblebee hive

The number of foragers working from each colony ranged from one to forty-five, with a mean of eighteen. The trapping vestibules did not account for the foraging work of males, which leave the colony and do not return, but they actively forage to maintain themselves. I had no way to estimate the male production of individual colonies, but the flower observations showed around 30% as many males as workers on the flowers. Thus the effective foraging strength measured by the trapping vestibules should be increased by 30%, which would take it up to twenty-three foragers per hive.

The bumblebee hives placed on the crops were only intended to create a range of bumblebee densities, not as an experimental treatment. However, it was interesting to estimate their contribution to the numbers seen on the crops, i.e. how many of the B. ruderatus foragers were 'mine' and how many were from the wild population. The table below compares the total number of long-tongued foragers on the four crops (from the number per square metre in the test plots extrapolated to the crop area) with the number of foragers contributed by the added hives (from the mean number of foragers trapped per hive, multiplied by the number of hives on the crop).

Crop ID	A	B	C	D
Area (heactares)	0.065	0.71	1.2*	1.5*
Number of colonies added	19	56	70	0
Total Long-tongued bees from counts on crop.	357	2464	1938	807
Total long tongued bees estimated from hive traps	437	1288	1610	0

Table of long-tongued bumblebees on four crops. **On the larger crops a significant proportion of the long-tongued bumblebees could be attributed to the added hives. The smallest crop was apparently too small to accommodate all the bumblebees placed there and can be considered to be saturated with pollinators. (* indicates area was estimated.)**

The nineteen hives placed on the smallest crop appear to have generated more foragers than were seen on the crop. This is not surprising given the very high hive density: nearly three hundred hives per hectare (125 per acre). We can assume the crop was saturated with pollinators, and it is likely that some dispersed to crop

B, which was well within flight range. On crops B and C, the added hives were estimated to contribute 50% and 80% of the total foragers, respectively.

Number of bumblebee hives needed

Despite none of the crops reaching more than 63% of florets setting seed, the trend with increasing bumblebee density suggests pollination was reaching saturation. Three thousand long-tongued foragers per hectare (1250 per acre) apparently achieved maximum pollination.

(hives per hectare) = (foragers per hectare) / (foragers per hive)
(hives per hectare) = **3000 / 23** = **130 hives**

From the formula above, 130 hives per hectare (55 per acre) of B. *ruderatus* would be required to give optimum pollination of tetraploid red clover. This assumes they are the only pollinators present, which is unlikely to be the case, but it is a useful figure for estimating economic value.

The above calculation is based on the bumblebee density that appears to approach saturation of seed set based on flower visitation and seed numbers from one cohort of flowers. Another approach is to relate the bumblebee density to the weight of harvestable seed from the total harvest. This will represent a wider cohort of flowers, and correct for lost seed. Hand-harvested plants from the 2 x 2m plots, processed with a threshing machine, yielded a mean weight of 241 g (0.53 pounds) per plot, which had a mean long-tongued bumblebee density of 1.8 long-tongued foragers per plot. So if 1.8 foragers produce 241 g seed, the 23 foragers per hive should produce 23/1.8 x 241 g = 3.1 kg (7 pounds) of seed.

The similarities and differences between my results and others are notable. My estimate of 3000 long-tongued foragers per hectare is comparable to the 4000 /ha estimated by Macfarlane *et al.*, but my estimates of the number of foragers and the amount of seed set per bumblebee colony are far smaller (23 vs over 200 foragers, and 134g seed per forager vs 600g).[15]

Would commercially produced B. *ruderatus* colonies be economically viable?

Red clover seed prices have changed over time, but $6/kg is reasonable. The best crops in this trial produced an estimated 700 kg/ha (620lb/ac), which would be worth $4200. If we accept my estimate of 130 hives per hectare, that is $32 of seed per hive. Obviously, pollination is only one of many costs of growing the crop, so the hive price would need to be a fraction of the $32. The alternative calculation of 3.1 kg of seed per hive comes to a value of 3.1 x $6 = $18.60.

The short answer is that the commercial supply of hives of long-tongued bumblebees for red clover pollination is unlikely to be viable. This leaves little alternative to low-cost habitat modification as a method of increasing the numbers of long-tongued bumblebees around red clover seed crops. Macfarlane *et al.* gave the example of a farmer who grew broad beans as an early crop and set aside an un-cut section of red clover to produce flowers after the beans and before the seed crop. This was in conjunction with setting out sixty wooden domiciles, of which twenty-one were occupied by *B. ruderatus*.[16]

Kiwifruit

The baby boomer generation of New Zealanders remember the 'Chinese gooseberry' as a brown furry fruit, growing on untidy vines in a few back gardens, and how they were rather pretty cut in cross sections with their ring of black seeds in the green flesh. The story of their re-birth as 'kiwifruit' and their explosion onto the international market has been told elsewhere.[17] Suffice to say, by the 1980s fortunes were being made, especially from land sales and speculation, and the industry was expanding rapidly. (I deal more with the industry issues and research funding in the next chapter.) With kiwifruit being very pollination-dependent, there were serious concerns about the ability of the honeybee industry to continue to be able to supply enough hives, and those of us with experience with bumblebees felt they should be considered.

Size matters for most fruit crops. For given levels of nutrients, water and light, fruit size is mainly driven by pollination and fruit loading (how many fruit the plant needs to 'feed'[18]). Apples and some other fruit crops produce

Figure 3.12 Kiwifruit (green, cultivar Hayward) on the vine.
Photo credit: Jocelyn Winwood.
Figure 3.13 Kiwifruit (gold, G3) in cross and long section.

far more flowers than can develop into fruit of an acceptable size, so size is managed partly by thinning out a proportion of fruit while still small. Kiwifruit vines produce fewer flowers than apple, but some thinning is often practised. Hormones from the seeds stimulate fruit growth, and the number of seeds, in turn is the result of the number of pollen grains that reach the receptive parts of the flower and go on to fertilise the ovules at the base.[19] Green kiwifruit (Hayward cultivar) require 600 – 1000 seeds for an export-sized fruit. The 'gold' cultivars require about half as many seeds.[20]

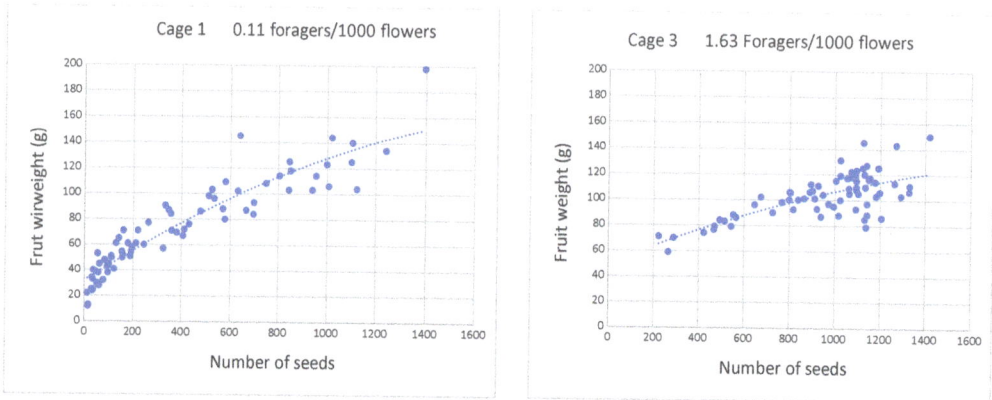

Figure 3.14 Graphs of kiwifruit size vs seed number (green, Hayward cultivar). This was a subsample of fruit selected to cover the full range of weights. Note the difference between the two cages presumably attributable to fruit loading: where there was better pollination and larger fruit, more seeds were needed to form the same sized fruit. 1000 seeds in cage 1 produced a 130 g fruit compared to 100 g in cage 3.

Kiwifruit pollination poses particular challenges. Fruit are borne on 'female' vines and must be pollinated from non-fruiting 'male' vines.[21] Neither male nor female flowers reward pollinators with nectar, but bees collect pollen from both types. Male flowers produce vast amounts of dry pollen, which can blow around in the wind, as with grasses and pine trees. Female flowers have large multi-armed stigmas, which can intercept wind-blown pollen, and are unlike the typical single-spike stigmas of most insect-pollinated flowers. But whereas wind-pollinated plants typically have drab flowers that lack visual cues, kiwifruit flowers have large, showy petals, not unlike wild roses. Female flowers have stamens that produce pollen, but the grains lack cell contents, and it is considered quite deficient as a bee food, but it is collected by both honeybees and bumblebees.[22]

Honeybees have always been regarded as the primary mode of pollination for kiwifruit. But the late spring flowering time often overlaps with white clover bloom in the surrounding countryside and even under the vines, and honeybees often find this a more rewarding forage source. Beekeepers minimise this by providing extra sugar-water in the hives, so the colonies' primary demand is for pollen. A study with various beehive stocking rates recommended eight hives per hectare (3.3 hives per acre).[23] This was at least twice as many hives as were usually recommended for other orchard crops. Subsequent work by Mark Goodwin indicated that it did not make much difference how many beehives were placed in the kiwifruit orchard.[24] The bees spread over the local countryside, and it was more a question of how many bees were needed to saturate the surrounding clover or other forage, to make the pickings lean enough that kiwifruit rose up their priority list.

Figure 3.15 *B. terrestris* workers on kiwifruit flowers. **Left: Female flower. Note multiple white stigmas below the bee; Right: Male flower with abundant pollen-yielding stamens but no female parts.**

Design of the trials

As explained in the red clover section, the ideal outcome of pollinator stocking-rate research should be a dose-response curve, i.e. crop yield versus the number of pollinators, derived under controlled conditions. The red clover trial was limited by being uncontrolled, but it did show a trend between pollinator numbers and seed yield. For kiwifruit, I hoped to create a properly controlled experiment. This usually involves a compromise between realism and certainty.

In a properly controlled experiment, the manipulated factor (e.g. the number of bumblebees) should be the sole source of variation in the output (e.g. yield of kiwifruit). The only practical way to manipulate numbers of bumblebees was to confine them to the kiwifruit vines with cages. This would not have been possible with honeybees because the colonies are too populous to confine in practical-sized cages and achieve realistic bee densities or normal behaviour, but bumblebee colonies are much smaller. My idea was to enclose a whole female vine and part of a male vine to create a microcosm of an orchard. I could not just shove different numbers of bumblebees into each cage: workers only forage for pollen if working for a colony, so I needed intact colonies. But they needed to be very small colonies, or their numbers would have been unrealistically high for the size of the cages. Placing a colony with, say, twenty workers in a cage still did not tell me how many foragers would be working the flowers, so this was monitored directly by regular counts of foraging bees, as I did with the red clover. I built one set of four cages, each about 5m square, in a Levin orchard, which was near enough to the University for day-trips. A recent graduate, Gail Wilson, went to Te Puke, the main kiwifruit growing area, to manage data collection in another four cages. The flowering period was less than two weeks, after which the cages were dismantled, allowing the fruit to develop under open-orchard conditions.

I was aiming for 0, 1, 2, or 4 foragers working in the four cages. The number of bumblebee workers we counted foraging on the flowers in the cages turned out to be less than half the number I was aiming for. I repeated the trial nearly twenty years later with much larger cages at Te Puna, with higher worker densities, although again, the foragers counted were fewer than my target (that later trial is mentioned in more detail in Chapter 7).

Figure 3.16 Cages in a kiwifruit orchard. Shows one cage and part of another, each enclosing one male and 3 – 3.5 female vines. Dimensions 3.8 m wide, 2.5 m high and 25 – 30 m long (12 x 8 x 80 to 100 ft).

Results: Kiwifruit seeds versus bumblebee density

Although fruit *size* is the fundamental economic factor, size is driven by the number of seeds and varies with fruit loading (see figure 3.14). Seed number is the most direct measure of pollination. On figure 3.17, each dot is the mean bumblebee density and seed number for a single cage. The green dots are from the 1982 trial at Levin and Te Puke, and the brown dots are from the much later trial in 2000, the background of which is explained in Chapter six. My metric of total foragers (on *both* male and female flowers) per thousand *female* flowers may seem strange, but my logic was thus: only the female-visiting bees are actually setting fruit but in terms of the number of foragers working from a hive, I needed to account for those present on both types of flowers. Of course, it is primarily foragers *changing* from male to female flowers that are effecting pollination. The 2000 trial showed overlap with the earlier results. However, even the cages with the highest density of foraging bumblebees were only just in the optimal export-size range of fruit set. We clearly had not reached saturation point.

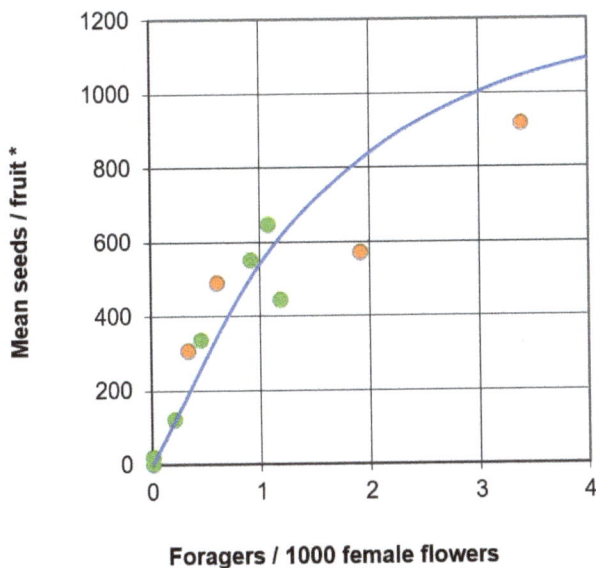

Figure 3.17 Graph of kiwifruit seed number vs number of foraging bumblebees. Each point represents data from one cage: Two orchards in 1982 (green), one orchard in 2000 (brown). Number of foragers includes those on both male and female flowers. * Mean seeds are estimated by deriving a seed number vs fruit weight equation from a representative sample from each cage, and then calculating the presumed seed number to all fruit. The line is based on an asymptotic relationship with maximum = 1200.

From the 1980s trial, I concluded that to produce export-sized kiwifruit, about two *B. terrestris* foragers per 1000 female flowers were needed. When I extended the work in 2000, I updated that to around four per 1000 flowers. Based on 420,000 female flowers per hectare, that equates to 1700 foraging bumblebees per hectare (174,000 flowers, 700 foragers, per acre). Work I assisted with in 2017 (using different methods) on 'gold' kiwifruit suggested 914 foraging bumblebees per hectare (380 per acre), which is consistent with the estimate of 1700 for green 'Hayward' considering the gold cultivar requires about half the number of seeds to set a commercial-sized fruit.[25]

Results: Bumblebee foragers per colony

The colonies in the cages were too small for their foraging strengths to be relevant to the full-sized colonies that would be used for open-orchard pollination. So, as with the red clover trials, we needed to measure the foraging strength of *B. terrestris* colonies. In subsequent seasons my group used trapping vestibules in open orchards, and the number of foragers per colony ranged from around five to seventy, but they were not all collecting kiwifruit pollen.[26] During my 2000 trial, all the hives had built-in trapping vestibules (Figure 6.1), and I recorded a mean of 23 foragers per hive, of which seventeen (74%) were carrying kiwifruit pollen. A later trial by Cutting *et al.* arrived at a similar number of foragers, but it should be said that in both their trial and my 2000 trial, the colonies were known to have aged past their peak strength by kiwifruit flowering time.[27] The following graph exemplifies the variability and rapid change in bumblebee foraging strength over time.

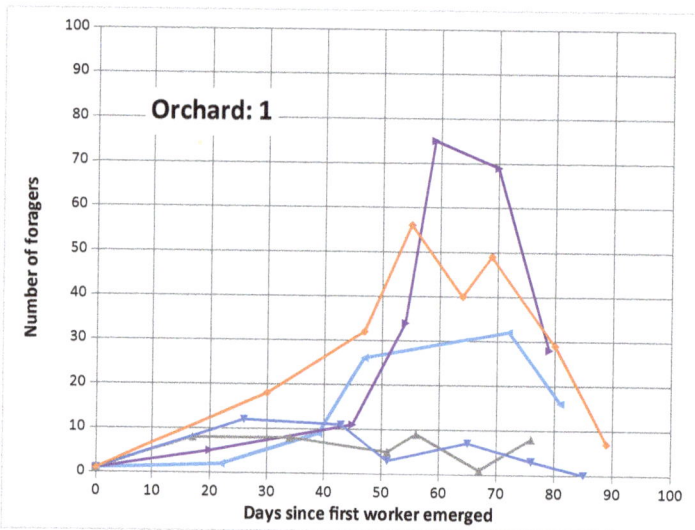

Figure 3.18 Foraging strength over time of five *B. terrestris* colonies in a plum orchard. **The forager counts were taken from trapping vestibules at approximately weekly intervals. Note the extreme variability between colonies.**

Estimating stocking rates

I settled on a figure of twenty-five kiwifruit foragers per hive as a basis for calculations. If 1700 foragers per hectare were needed for green (Hayward) kiwifruit, the number of hives would be 1700 / 25, which equals 68 hives per hectare (28 per acre). So I concluded that approximately seventy bumblebee hives would be needed per hectare as the sole pollinator. The 2017 trial on 'gold' kiwifruit indicated 914 foragers per hectare (380 per acre) which suggests 37 hives per hectare.

The above stocking rates are much higher than the figures recommended by commercial bumblebee suppliers. In New Zealand, Biobees recommend fifteen 'turbo hives' (approximately 200 workers each) per hectare of kiwifruit in the absence of honeybees, and Zonda Beneficials recommend at least nine 'large hives' (80-110 workers) per hectare.[28] These recommendations appear to be following advice from European associates: Biobest report that twelve hives per hectare (minimum 200 workers per hive) provide full pollination. This was based on a trial in a caged section of an orchard in France in collaboration with Zespri Group Ltd.[29] Why is there such a big discrepancy? The unpublished Biobest/Zespri trial appears robust although there are no data on the number of foragers working per hive or the density of foragers on the flowers, which makes it difficult to compare with my data.[30] I suspect the European commercial bumblebee colonies are larger than the New Zealand ones (see discussion of the different subspecies in chapter 8). More direct measurements of foraging rates are needed.

The foregoing chapter has included data and conclusions from later times for consistency of subject matter. But before we had a clear estimate of the required bumblebee stocking rates, we assumed it would be 'a lot'. So we urgently needed to find a way to rear huge numbers of bumblebees economically. That is the story of the next chapter.

Nelson Pomeroy **Bumblebee Keeper**

Chapter 4

Towards bumblebee domestication

Towards bumblebee domestication

Early during the kiwifruit pollination work (1982), I gave a seminar at Massey University about the possibility of domesticating bumblebees for crop pollination. A gentleman from the Development Finance Corporation (DFC) attended my seminar and enthusiastically encouraged the bumblebee programme. The DFC had loaned large sums to new entrants to the kiwifruit industry and had a stake in its success. A DFC study had extrapolated the recent growth rates of the kiwifruit and the honeybee industries, and had estimated that by 1986 the demand for hives of honeybees to pollinate kiwifruit would outstrip the supply. In an informal moment of largesse, he asked me, 'If we gave you a quarter of a million dollars, what could you do?' I was stunned. At this stage, I was a junior academic on a short-term contract. What did I do? I planned a new system, with (short-tongued) bumblebee colonies produced in an indoor rearing facility, 'fattened up' on a farm with wall-to-wall flowers of types rich in nectar and pollen, and freighted by the truckload to kiwifruit at flowering time. After kiwifruit flowering, the hives would be retrieved, and progeny queens taken for following year's breeding stock. I priced it up at half a million dollars, including a two-hundred-hectare (480 acre) farm, buildings, and a three-tonne truck. While poring over my archives for writing this book, I found the cover page of the proposal: *Kiwifruit Pollination by Mass Produced Bumble Bees: Proposal for a new technology. N Pomeroy and BP Springett, Department of Botany and Zoology, Massey University, June 1982*. (Brian Springett was the Professor of Zoology and my Head of Department.) It turned out that sober heads in the DFC felt my proposal was premature, but Wrightson Horticulture, with whom I had good connections following their sister company Wrightson NMA's support for the red clover work, were fully supportive, albeit at a more modest scale. They, along with the New Zealand Kiwifruit Authority and a private donor, supported the pollination trials described in the previous chapter, and the rearing work described in this one. Career-wise I was in a strange position. I was employed on a one-year (renewable) contract as a postdoctoral fellow – most post-docs did their time and moved on to a more secure job. I was the most junior academic in the department and soon received more research funding than anyone else. My fellowship was renewed a couple of times and then converted to a lectureship (equivalent to an assistant professor in North America).

At the time the big push came to consider bumblebees for kiwifruit pollination, I had not started the pollination trials described in the previous chapter, so we did not know how many bumblebee hives would be needed, but nationwide it was bound to be 'a lot' so a way needed to be found to domesticate bumblebees on a commercial scale.

Proposed production cycle

Each step in the bumblebee life cycle had been achieved in captivity over the years by several researchers, but the stage that seemed most labour-intensive and skilled was the initiation of colonies from individual queens.[1] Figure 4.1 shows a chart of the life cycle stages with respect to the rearing process and pollination. The big plus for using B. terrestris in New Zealand was that the species was abundant, and the natural peak of colony strength often occurred in late spring, when kiwifruit flowered. (I must keep stressing that even in cooler, shorter-season climates, bumblebee colonies are seldom terminated by the onset of winter – they reach maturity and decline much earlier for other reasons.)

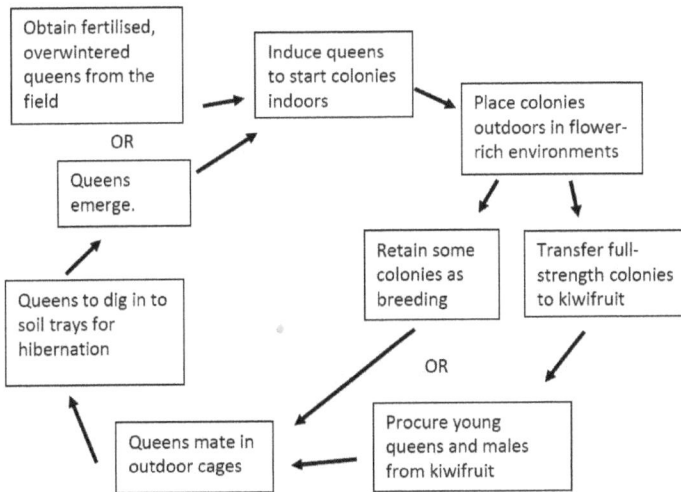

Figure 4.1 Flow diagram of the proposed rearing system for kiwifruit pollination.

Obtaining queens

Although I was fortunate to have access to a habitat where the long-tongued B. ruderatus was common and was suitable for red clover pollination, experience suggested there was no advantage to using the long-tongued species for pollinating kiwifruit: the flowers were shallow, the short-tongued B. terrestris was even more common in most districts, and it appeared to have larger colonies. 'Overwintered' queens appeared in late winter (July) in warmer regions.

New Zealand is sometimes regarded as a 'biodiversity hot spot' due to a large proportion of the native plants and animals being unique to these islands.[2] These species are mostly strictly protected by law (except for most fish, an ecological anomaly, but understandable culturally and common to many countries). But many of the familiar species have been introduced, by accident or design, from other land masses and have no such protection either in law or in popular opinion. Honeybees have some statutory protection in terms of prohibiting insecticide use on flowering crops, but this arose as a property-protection move, as beekeepers owned the bees, and the industry demanded protection of their stock in the same way as it is unlawful to harm or steal other people's sheep and cattle. Bumblebees were just another exotic species. If they were introduced now, they would be regarded as an invasive species, as they were when they first appeared on the Australian island of Tasmania in the 1990s (more on that in Chapter 8).

We anticipated that many thousands of queens would need to be collected from the field over a short time in early spring to initiate enough colonies to make a real contribution to the kiwifruit industry, at least until we could breed our own queens. I thought this would be fine due to the natural abundance of queens, and indeed later, when we were sometimes collecting large numbers each day from a single location, we could detect no diminution in numbers over time. I discuss some broader aspects of bumblebee conservation in Chapter 8.

Queens were captured in nets, usually while foraging. The ideal forage source was tree lucerne, also known as tagasaste (*Chamaecytisus palmensis*). This is an amazing plant introduced from the Canary Islands.[3] It mass-blooms over several weeks, and variation between plants provides forage from early winter to mid-spring. It is commonly grown for roadside bank stabilisation, and there are many quiet country roads where you can park and collect queens from the bushes. Queens with corbicular pollen loads were ignored (or released if caught) as they would already have incipient colonies that we did not want to orphan.

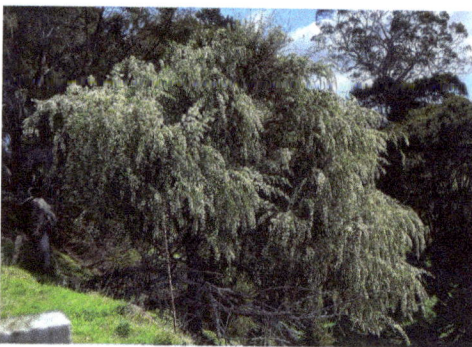

Figure 4.2 (above) Tree lucerne bush (*Chamaecytisus palmensis*).

Figure 4.3 (right) *B. terrestris* queen foraging on tree lucerne.

We developed special equipment for this scale of collecting, where queens were so abundant that the only limitation was the time it took to sweep the net and extract the queen. Our standard net had a two or three-piece telescopic handle, as used in window-washing brushes, and could extend to about three metres (10 ft). We designed net bags with a screw thread at the tip so a queen-holding tube could be screwed into the net. With a firm sweep, the netted bumblebee tumbled into the tube. The net was usually flicked down to the ground to prevent escape, the queen-containing tube was unscrewed, a cap applied, the occupied tube placed in a bag, and a replacement tube screwed onto the net. If more than an hour would elapse before returning to refrigeration facilities, the queens were stored in insulated bins with ice packs.

When we collected more queens in a day than we had time to place in the colony-starting apparatus, we fed them and held them in cool storage. We vacuum-formed trays that contained rows of depressions into which the queen tubes could rest and long narrow troughs to hold liquid food. The trays matched the tube slits so that queens could be lined up in rows with one slit resting a few millimetres above the liquid trough. The queens stuck their tongues through the slits and drank the food. When fully fed – usually discerned from the queen settling in one place with rapid abdominal breathing motion and responding with a high-pitched buzz if disturbed – they were stored in a fridge for up to several days before installation (around 4°C, 40°F).

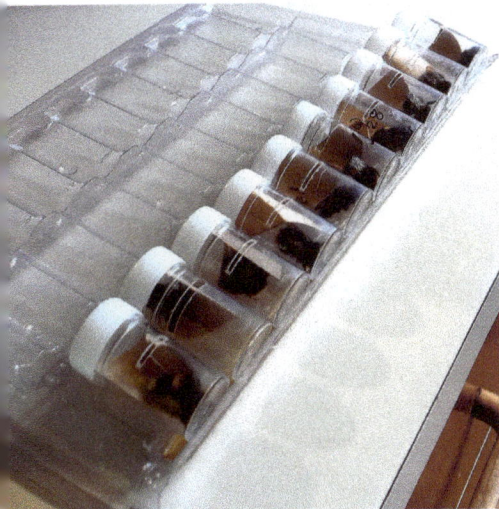

Figure 4.5 *B. hortorum* queens feeding. *B. terrestris* have shorter tongues but can also feed through the tubes' slots.

Figure 4.4 feeding tray. The custom-cast tray has thirty recesses for placing specimen tubes of queens. Each row has a channel for feeding solution.

Starting colonies

In the field, bumblebee queens always start their colonies in an accumulation of fibrous material that provides thermal insulation. They collect pollen which they store as a small lump, on which they lay their first eggs, which they cover with wax. Adjacent to the pollen lump, they fashion a wax pot into which they regurgitate nectar after foraging. They unload pollen from their hind legs against the side of the developing brood clump. The natural habit of starting colonies inside fibrous nesting material enables queens to maintain a high temperature there. She straddles the brood clump and applies body warmth to it just like an incubating bird.[4]

Most previous methods of bumblebee colony initiation had involved supplying a lump of pollen, either within a cavity in fibrous material or in a warmed environment. In Chapter 2, I mentioned that in Chris Plowright's group in Toronto, we routinely initiated colonies of a range of species. Queens laid their first eggs on a small disc of pollen dough inside a cavity of upholsterer's cotton.[5] The thermal insulation of the cotton wadding enabled the queen to keep her nest cavity warmed to around 30°C (86°F), and the surrounding room temperature was not very important. The big drawback of this system was the amount of work needed to uncover the queen's cavity every day or two to change the pollen lump until she laid eggs on it, and thereafter to provide fresh pollen and monitor progress. These activities disturbed the queen and required skill not to mess up the cavity. The alternative method without nest material either involved warming a room to 30°C (86°F), which I felt was wasteful of energy and an unpleasant place to work, or placing the colony-starting containers in incubators, which was cumbersome in terms of handling and disturbance.

We were happy in Toronto that one person could look after about two hundred colony-starting queens, but I was aiming for one person to look after two thousand. I wanted a new system where the queen would lay her first eggs on a predictable spot, didn't rely on pollen lumps or nest material and was spatially compact. I also wanted to minimise the amount and skill level of labour and cost of equipment. This reads like an ambitious list now, but at the time, I just wanted it to work, and economies could be gained later.

In the field, *B. terrestris* queens readily invade and attempt to usurp existing colonies. If successful, the invader's first eggs will be laid on existing cocoons, not on a pollen lump (as are all post-incipient eggs in normal colony development). Therefore, an imitation cocoon might be an attractive surface on which a confined queen might lay her first eggs. Thus began trials with a convex, warmed knob, scented with bumblebee wax.

The sloping sides of the conical observation hives (page 36) were originally intended to support peripheral brood clumps and retain a compact comb shape. The same reasoning suggested using a rounded bowl or cup shape for the colony-starting containers. But I thought it would have additional advantages: it might tend to focus the queen's activity towards the centre, and pollen could be imprecisely dumped anywhere, and it would roll to the centre, beside the knob and the brood clump thereon.

Compact box version

Initially, I used plywood boxes with sliding acrylic lids but made the feeding and defecation chamber much smaller than we had used in Toronto. To create a bowl shape for the nesting compartment, I inserted a block of plaster of Paris cast to that shape, with a central hole for the knob, and a small pit (lined with wax) as a template for the queen to make her nectar pot.

Figure 4.6 Compact wooden starter boxes. **The brood compartment included a plaster block moulded to provide a bowl shape. A central knob was warmed from below.**

The first knobs were thimble-sized hollow domes made from two layers of stretchy T-shirt cotton, sandwiching a layer of polythene plastic. The sandwich was heated with a clothes iron to soften the plastic and then press-moulded into shape. My students dubbed these 'nipple warmers'. The knobs, and to some extent the surrounding plaster blocks, were heated from below with ceramic resistors epoxy-glued to small squares of sheet steel (25 x 25 mm, 1 x 1 inch). A row of boxes was aligned so that the knobs rested on a row of these miniature heat plates. Controlling the knob temperature directly with a thermostat would have had a lot of hysteresis (on-off time lag due to the slow heat conduction through the knob), so I placed the whole row on a home-made 'Simmerstat', which I adjusted to achieve a knob temperature of 30°C (86°F).[6]

Plastic starter cup system

I would have liked to have made the entire containers from a porous material like the plaster insert, and I did experiment with a plaster/cloth mixture. However, I could not find any material that lent itself to large-scale production and was compatible with a simple transparent lid. Bryan Wenmoth of Massey's Production Technology Department suggested I should try plastic. He explained how thin plastic sheets could be heat-softened, and shaped over a mould by being drawn down with a vacuum, or by being pressed between 'male' and 'female' moulds. Such casting processes naturally left a flat flange on which a clear plastic lid could be slid, whose edges had been folded over after heating with a hot wire (like the cardboard and plastic retail blister packs with slide-on clear covers).

After queens had settled and laid their eggs, they didn't spend any more time in the outer compartment than they needed to do a quick poop and drink sugar-water, so I reduced the size further and placed the sugar-water reservoir outside the container. I had tried wicks before to provide sugar-water by capillary action, but they tended to dry out. But then I found a broken felt-tipped pen on the sidewalk and saw it had a plastic-sheathed internal wick that served as the ink reservoir. After soaking and squeezing out most of the ink, it turned out that sugar-water would conduct up the shaft, and we could buy blank wicks from a pen manufacturing company in Dunedin. The wicks were only 5 mm (3/16 inch) in diameter and could be poked through a hole in the outer compartment. Queens licked the sugar-water off the end. The jar and wick could be replaced without opening the cup, which further simplified the care of the queens.

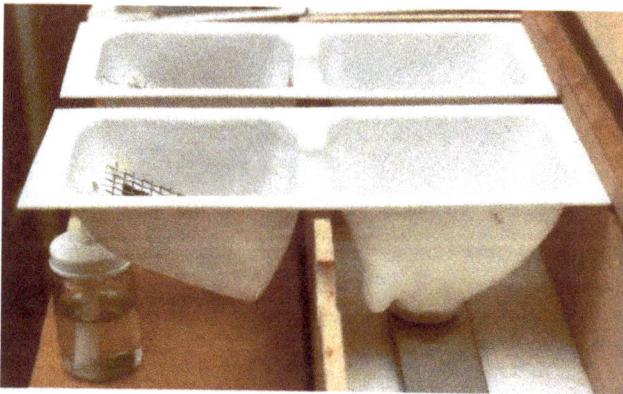

Figure 4.7 Early-version starter cups. **These were hand pressed from heated 1 mm high-impact styrene. They rest on a warmed metal strip.**

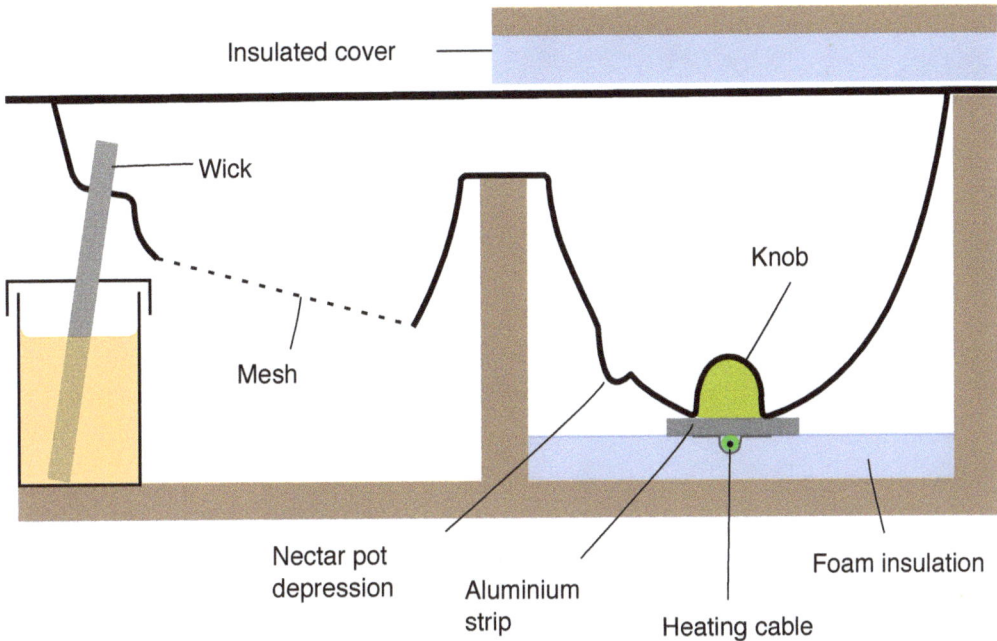

Figure 4.8 Longitudinal section through starter cup. The hollow knob is filled with Plasticine for heat conduction.

The knobs in the wooden box version had a soft fabric surface which I thought would be attractive to queens, but some were shredded, and the low thermal conductivity made them less suitable as a 'warm spot'. So subsequent knobs were smooth-surfaced. The earliest starter cups had a small flat area in the centre of the brood compartment, onto which was stuck a solid knob (cast from wax-impregnated plaster of Paris). All later versions were like Figure 4.8, the knob shape was cast into the plastic and was filled from below with solid material (usually Plasticine modelling clay).

I was always concerned about the impervious smooth plastic surface, so we tried to make it more rough and absorbent. Could we laminate a thin stretchy layer of cloth to the plastic? Kate, a technician employed on the project, donated a couple of pairs of pantyhose. We slid an unformed rectangle of plastic into the leg of the hose and sprayed it with aerosol 'diesel engine start', which contains the powerful solvent ether. This dissolved the surface of the plastic enough to bond the nylon mesh to it. The mesh was sufficiently stretchy and heat-resistant that after the solvent had dried, the laminated piece could be heat-shaped, much like plain plastic. The mesh-coated cups provided grip for the queens to climb up the sloping surfaces easily, but did little to absorb faecal moisture, and the embellishment was not pursued. We subsequently provided a

Figure 4.9 *B. terrestris* queen with her first brood in a starter cup.

A. Nectar pot has been made by building a wax rim over the moulded
   hollow. Eggs have been laid at the lower left of the knob.
B. Half-grown larvae on left and pollen granules on right.
C. First larval group have made cocoons, and a new row
   of egg cells have been made on the knob.
D. First worker has emerged.

strip of matt-surface sticky tape to provide grip for the queen to climb between
the chambers.

I wanted the central knob to be the warmest spot, but the whole nesting
chamber also needed to be warm (high 20s °C [70s °F], I thought) without overheating the room. At first, I cast knobs from polyester resin with embedded

resistors. As with the wooden box version, the resin knobs would be fixed to the shelf, and the cups fitted over them. These seemed too tedious to make on a big scale and required numerous wiring connections. The large-scale version used a strip of aluminium underneath the cups, the only contact surface being under the knob. This would ensure that the knob was the warmest part and that the excess heat radiating and convecting off the rest of the strip would have a moderate warming effect on the whole brood compartment of the cup. The heat source at Massey in the 1980s and later at Zonda in the 1990s was soil-warming cable which came in a fixed length, ran off 240v AC and included an earthing sheath inside the electrical insulation. Other applications have used twelve-volt systems. The cable was either fastened to a flat bar (25 x 3 mm (1 x 1/8 inch) section) with heat-conducting aluminium tape or pressed into the slot in a sheet-joiner profile and embedded there with heat-proof stove putty. Our cables were 6 m (20 ft) long, were attached to three 1.8 m (6 ft) lengths of aluminium, and were arranged to heat three parallel rows of starter cups with 22 cups per row. A digital thermostat sensor was attached to the metal strip.[7]

The starter cups did not need to be opened to replace the sugar-water, and likewise, pollen was dispensed through a V-shaped cut in the flexible clear plastic roof. Pollen, and the queens when they were first installed, could be pushed through the slits, which acted as self-closing trapdoors.

The first cups (Figure 4.7) were hand pressed between male and female moulds. My long-term technician, Stan Stoklosinski, hand carved the shape from pieces of wood and dowel – half-round for the channel between the compartments, and a small rounded piece for the nectar-pot bulge. The female mould was cast over the waxed wooden shape with polyester 'bog' (as used in car panel repair) and set into a wooden frame. A rectangle of plastic (1 mm [0.04 in] high-impact styrene) was clamped over the female mould, heated with a domestic radiant heater, and the male mould was pressed into it. This technique is reasonably practical for anyone wanting to make these units on a small scale. Once we were in the scale of hundreds, a local small-scale vacuum moulding specialist, Nolan Plastics, made the units in 0.5 mm clear PVC over cast aluminium moulds. More recently, I have acquired my own machine and vacuum formed the units shown in Figure 4.9 (more on this project in Chapter 7).

Colony expansion outdoors

The colony-starting system took in single queens and husbanded them up to the emergence of the first batch of workers. From that stage, it would take several weeks for each colony to grow to full strength for pollinating the crop. In Chris Plowright's lab in Toronto, we routinely maintained bumblebee colonies indoors, artificially fed (on sugar-water and honeybee-collected pollen) throughout their colony cycle. Still, I was not in favour of this for our mass production – I thought it would be too expensive in pollen, facilities and labour, and I believed continuously confined colonies would become too crowded, and would fail to grow as large as free-foraging ones.[8] So I elected to place the colonies outdoors to be self-maintaining.

In nature, the queen continues to leave the nest to gather nectar and pollen even after the first few workers emerge, but as more workers accumulate and become strong enough to fly, the queen can stay in the nest. I assumed our confined queens might get lost if allowed to leave the nest, so we waited until around ten workers had emerged from the cocoons. Then the incipient colonies were placed in larger outdoor hives and covered with nesting material: either upholsterer's cotton or fibreglass household insulation.

The outdoor hives were un-elaborate compared to the domiciles or starter cups – they did not need to attract queens or stimulate nesting – they just needed to provide space for expansion and adequate weather protection, ventilation and drainage. We used cardboard cartons (sometimes wax-impregnated) with a builders' foil rain cover, or lidded ten-litre plastic buckets with horizontal ventilation slits (cut with a table saw). The buckets were lined with lift-out cylindrical cages of plastic mesh that contained the colonies.

So, exploiting the University's rural land space, and in particular the enthusiasm and expertise of the Agronomy Department, we grew large plots of a variety of flowering 'crops' on which we placed our colonies to grow. John Baker had developed an innovative no-till seed-drilling machine and was happy to do test runs with diverse seeds from garden borage to blue lupins. After two seasons on a small patch at the edge of campus, we went to a larger scale on Keeble Farm with fodder radish, kale and other crops.

Figure 4.10 Flower field on Massey University campus. **From left: Tick beans (scarcely flowering), blue lupin, broccoli or kale, fodder radish, borage.**

Food resources likely limited bumblebee populations, and it seemed evident that you could not place unlimited numbers of bumblebee colonies outdoors and assume they would find enough nectar and pollen. So the questions were: how much nectar and pollen does a bumblebee colony need, and what quantity of what type of flowers would be needed to support each colony? My PhD student, David Woodward, investigated both questions. Taking into account his results under various conditions, data from MSc student Colin Tod's experiments and some earlier calculations I'd made with *B. ruderatus*, it looked like a typical bumblebee colony would consume around 100 grams (3.5 oz) of fresh pollen during its development.[9]

David studied borage *Borago officinalis* and three types of crucifers as likely candidates that could be easily grown, and that bumblebees liked to forage on. His measurements were exhaustive on flower density, nectar and pollen per flower, rate of new flower opening and the effects of weather and competition from honeybees. The most promising crucifer was fodder radish. Putting together the data from our various sources, it seemed a reasonable estimate that one- to two-hundred colonies of bumblebees could feed themselves from a hectare planted in a combination of borage and fodder radish (40–80 on an acre). Other more tree-like flower sources, such as tree lucerne, were also candidates. We also considered that if nectar was limiting, sugar-water might be able to be provided. To save the labour of adding it to each hive, I considered making central feeding stations that honeybees could not exploit. I have experimented with this idea over the years, but given that many *B. terrestris* workers have tongues just as short as honeybees', it is not clear how one would design such an exclusion mechanism.

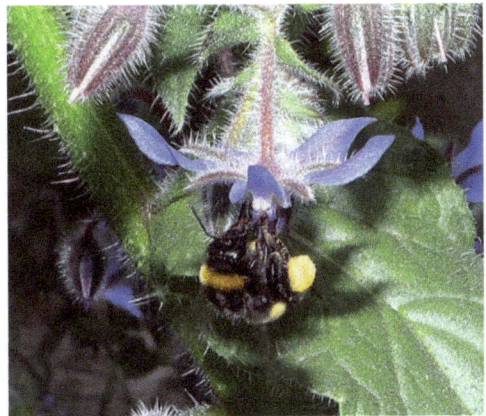

Figure 4.11 *B. terrestris* worker onflower of borage (*Borago officinalis*).

Transfer to kiwifruit orchards

Whereas honeybees are usually only moved at night to avoid losing foragers, bumblebee hives can be fitted with one-way entrances set at any time of day, and all foragers will be home within an hour. The hives are small enough to be lifted with one hand and can be kept closed for several days if supplementary food is inserted. Likewise, an untrained person can distribute and open the bumblebee hives at the orchard. They need a sugar supply due to kiwifruit lacking nectar, but a reservoir of less than one litre is enough to sustain them through the pollination period.

Bumblebee colonies are at their maximum pollen-collecting activity when rearing new queens at the climax of the cycle, so ideally this should coincide with the timing of crop pollination. After pollination, three options were envisaged: 1) If the colonies were in low-cost cardboard hives, let them live out their lives in or near the orchard; 2) Retrieve the hives after the colonies had naturally died out, at a time convenient to the orchardist, to recycle the more expensive, durable components of the hive; or 3) Retrieve the hives promptly, to make use of the new queens emerging in them for the following season.

Queen mating

Bumblebees lend themselves more than honeybees to total domestication because they mate readily in small spaces. Any well-lit cage is acceptable. We used walk-in cages under fluorescent tube lights in Chris Plowright's lab. Some researchers have been wary of the 50 Hz flicker of such lights because bees have 'fast' eyes and can detect flicker frequencies beyond 100 Hz.[10] We built a large walk-in mesh cage beside the Biology building and released young queens and males into it. Sometimes people noticed the sweet scent of the males' sex attractant pheromone from the other side of the building.

Figure 4.12 *B. impatiens* mating. Note that the queen's sting is fully extended and ends just under the male's chin. Photo credit: Jennifer Forman Orth.

Hibernation

I had tried overwintering queens of *B. ruderatus* back when I was starting my MSc research. I released males and new queens from observation hives into the old school room (shown in Figure 1.3) and provided large shallow bins (about 1 x 1 x 0.2 m, 36 x 36 x 8") of damp peat and also clumps of the spongy base of pampas grass (*Cortaderia selloana*) which had formed dense stands near our house. Many queens dug in to the peat and the pampas clumps, and I bundled the materials together in a large sack (farm wool pack) for transfer back to Massey University, but I don't recall that I managed to get any to survive and emerge.

In Chris Plowright's lab, the process was more organised. Mated queens were allowed to dig themselves in to shallow boxes ('flats' commonly used for germinating seeds) of compressed damp peat. They usually did this within two weeks of mating. Often it was 25°C (nearly 80°F) in the cages, and the light cycle was not controlled, so there was no 'Autumn' seasonal stimulus for digging in. At Massey, too, we provided boxes of peat and let the queens dig in, even though it was mid-summer.

I bought four domestic fridges – much cheaper than laboratory-grade chilled cabinets – and wired in more sophisticated thermostats and computer fans for thermal mixing. The trays of hibernating queens were wrapped in plastic rubbish bags to stop the peat drying and placed in a succession of fridges at a range of temperatures to step them down from the outdoor warmth to 5°C (40°F), where we kept them for several months. On warming again, they dug themselves out, but this would sometimes be quite delayed, especially if they were placed directly into 5°C rather than a more natural ramp-down. Other researchers found more reliable ways of waking queens from hibernation, which I will mention in Chapter 5.

Filing a patent, and disruption

We believed the starter cup system was world-leading as a method to initiate bumblebee colonies economically. It was duly patented by the University, with the commercial funders Wrightson Horticulture given the first option to commercialise the technology exclusively.[11] I assumed the colony expansion phase would occur outdoors, possibly on purpose-grown flower crops. Although I considered it possible to retrieve queen-producing colonies from kiwifruit orchards after flowering for mating and hibernation under controlled conditions, we had not refined this aspect of the production cycle on a large scale. It seemed quite practical to capture the queens from the field, at least during the early build-up of the proposed industry, and I doubted this would be practical outside New Zealand, where B. terrestris was abundant. Thus I advised that it would be unnecessary to get international patent protection.

Although we had 'cracked' the problem of large-scale initiation of bumblebee colonies from confined queens, by the end of the 1980s bumblebees had not been commercialised for kiwifruit pollination. This was partly because the urgency declined due to the industry managing satisfactorily with honeybees, and we had not developed all the aspects into a commercial package.

But then it all changed anyway. In the late 1980s the New Zealand kiwifruit industry was becoming fractured. Growers claimed that competition between the seven exporting companies was driving down prices, and legislation was enacted in 1988 that cancelled their exporting rights and required all kiwifruit to pass through a single marketing board.[12] Auckland Export, associates of Wrightson Horticulture, were one of the casualties, and Wrightsons promptly terminated all their funding of Massey's bumblebee programme. I could no longer pay Stan, my long-time technician, and other activities were curtailed. The University was owed a contract instalment of over $10,000. When the University requested payment, Wrightson responded by declining to pay but instead offering to forego all intellectual property (IP) rights.

To add to the instability, at the beginning of 1989, Rachel and I had moved to another town – Hastings, two hours drive away – while I still worked at the University Monday to Friday. This was a decision about our children's schooling, which I will elaborate on later in the next chapter. The arrangement was not going to be sustainable in the long term.

Chapter 5

Bumblebees in greenhouses – Genesis of a unique industry

Bumblebees in greenhouses – Genesis of a unique industry

While the number of bumblebee colonies needed to pollinate field crops like red clover looked prohibitively uneconomic, and we were still determining what would happen with kiwifruit, greenhouse crop pollination looked like a more practical use for bumblebees. Greenhouse crops tend to have a high value per plant or square metre; otherwise, they would not be economical in the face of the capital and energy costs of the greenhouses. Rick Fisher, a Canadian peer from my time in Toronto, was on a post-doctoral fellowship with my group and had been researching bumblebees' effectiveness in pollinating musk melons in a greenhouse. He estimated that one hive of *Bombus terrestris* may pollinate over five hundred melons worth over $1 each.[1] One hive yielding $500 of fruit was much better than the red clover seed estimate of $18 to $30. We had heard talk of using them for greenhouse tomatoes, and Rick placed a colony in a small tomato greenhouse. However, one of Massey's horticulture scientists suggested that as tomatoes were self-pollinating, bees would not be helpful.

But bumblebee pollination *had* been demonstrated to benefit tomato production in greenhouses. We were not aware that in 1987 a Belgium vet, Roland De Jonghe, who had had a long-time involvement with keeping bumblebees as an amateur enthusiast, had started raising them on a commercial scale for greenhouse tomato pollination and had founded the company Biobest.[2] Nearby, in the Netherlands, Koppert Biological Systems were established innovators in the mass-rearing of a range 'biological control' products, including insects and mites, to control greenhouse pests without chemicals.[3] They were seeking expertise on rearing bumblebees.

Koppert had approached a university in the Netherlands where world-class bumblebee research was being conducted but apparently could not obtain an exclusivity agreement preventing the intellectual property (IP) from passing into the public domain. Our programme at Massey University was already more focused on mass-rearing technology, and the administrators were willing to consider exclusive contracts. We soon began serious negotiations with Koppert. With a new funding partner, the problem with Wrightson went away: Massey accepted Wrightson's offer to forfeit the IP rights in return for forgoing payment of outstanding research funds. Koppert could take over both the IP and the funding.

Why tomatoes?

Tomatoes are self-pollinating. The pollen needs to move a few millimetres from the male to the female parts inside the flower. This sets the seeds, which in turn stimulate enlargement and ripening of the fruit. Outdoor tomatoes usually are moved around enough by the wind for this, and are a broad-scale, lower-value

Figure 5.1 *B. impatiens* worker on tomato flower. **The upper flower shows brown marks from bumblebee visits – an indicator of good pollination. Photo credit: Timothy J. Stanley**

crop where it is not so essential to achieve uniformly large fruit.[4] To obtain good yields of indoor tomatoes, the flower clusters were either shaken with a hand-held electric vibrator every few days (the norm in Western Europe) or sprayed with a hormone that mimicked the effect of seeds (more common in New Zealand), and which tended to result in seedless, watery tomatoes, especially during the colder months.

When bumblebees first became available for greenhouse tomato pollination in Europe, they were competitive with the labour cost of hand pollination but produced better yields of tomatoes.[5] And this was at a per-hive price of over US$100, much higher than we ever envisaged would be economical for the hive density needed for kiwifruit. Due to the self-pollinating nature of tomato flowers, a visiting bee did not need to be carrying compatible pollen; it just needed to shake the flower, so a single visit could produce a fruit. The recommended stocking rate was only six hives per hectare (2.5 per acre), a fraction of what I was envisaging for red clover or kiwifruit. Tomato crops, however, flowered for several months, much longer than the peak strength of a typical bumblebee colony, so colonies needed to be replaced a few times over the tomato season. These special features of tomato crops lend them to bumblebee pollination at prices and stocking rates profitable to both parties. In few other crops, especially outdoors, does bumblebee pollination even approach this level of cost-effectiveness.

Commercial negotiations

Massey University was limited in negotiating power over the starter-cup technology because they did not have patent protection outside New Zealand. Koppert understood this and could have 'run away' with the technology, but they wanted exclusivity and confidentiality. An agreement rewarded the University for Koppert's exclusive access to the technology and ongoing research. Massey University had a three-way profit-sharing arrangement with the relevant department, the individual staff member, and the University's central funds. The Botany and Zoology Department (soon to be restructured as the Ecology Department and the Plant Biology Departments) invested its share to create a scholarship fund, the J P Skipworth Scholarship, to support graduate students working in ecology.[6] My share was enough to make a big difference to the economics of our young family.

What followed was a whirlwind of activity. Director Henri Oosthoek and the company's bumblebee biologist Adriaan van Doorn (who had done similar PhD research to mine) made at least two visits to Massey over the next year, and I was flown to the Netherlands for several visits, some with a few days' notice and lasting only a few days. (There is a place on the globe further from Massey University than the Netherlands, but not by much: central Spain.)

Bumblebees were so cost-effective that all greenhouse tomato growers wanted them in the Netherlands and across western Europe. The demand was exploding. I was told that some French growers would get their wives to weep over the phone, pleading to be able to buy bumblebees. There was a trade-off between selling colonies or retaining them to use the emergent queens as new stock, and wild-caught queens became the basis for the early expansion to meet the demand. The favoured species was *B. terrestris*. Koppert were using queens from the Netherlands and other parts of Europe. This was also the commonest species in New Zealand (originally introduced from Britain), and we contributed to the early build-up of their breeding stock.

Exit from academia

At the end of Chapter 4, I mentioned that before Koppert became involved, Rachel and I had moved to another town two hours' drive away. This was not considered a reasonable daily commuting time in provincial New Zealand (and was unusual even in the biggest cities). To say the move was a 'complication' to my Massey career is an understatement. The most honest description is that we 'put it in the too-hard basket'. I kept a flat in Palmerston North and commuted weekly. Work hours were flexible enough to have Monday breakfast with my family, drive to Massey, and be back in time for Friday dinner.

We had moved so that our children (then aged two, six and ten) could attend the Rudolf Steiner school in Hastings. Rachel and I felt the prevailing public school culture lacked 'soul'. We weren't attracted to Christian churches, and she had become interested in what might be called esoteric philosophies. We felt the Rudolf Steiner schools offered a more 'whole person' education without religious dogma, and we joined a group trying to start a Steiner school near Palmerston North. I admit that my more hard-core scientific viewpoint clashed with some of what I read about Steiner's world-view, but I was impressed with Steiner teachers who visited our group and those whom I met at open days at the well-established school in Hastings. After a few years, we were convinced such a school wouldn't happen at Palmerston North (it still hasn't after thirty years), so we moved to Hastings. Our house was a hundred metres from the school. Our children were notoriously late as they sometimes ran out the door when they heard the bell for the first lesson.

Koppert had solved the first stalemate – by replacing Wrightson's funding – and they also helped solve my other problem of how to cope with a job two hours away from my home. They offered to fund me privately so I could quit the University and continue the research programme from our Hastings home. Some of my Massey colleagues certainly thought my exit to be foolhardy, and perhaps there was resentment for my terminating what had been a substantial contribution to the Department's funding.

Zonda Bees – a new company

It made sense to form a business entity for my dealings with Koppert and in anticipation of a bumblebee pollination industry in New Zealand. However, the greenhouse tomato industry here was tiny compared to Europe. For speed and simplicity, we bought a shelf company: literally 'off the shelf', i.e. that was already registered and ready to go but had not traded. Shelf companies usually have neutral names, and we chose one called 'Zonda Resources Limited'. 'Zonda' had a bit of a buzz sound – coincidentally, it was the local name of the hot, dry wind that blows down the Eastern slopes of the Andes. Shareholding was agreed at 50% Koppert, 25% Rachel and 25% me. Koppert's 50% input would be in funds, and our 50% would be in expertise.

The initial research was on queen quality and hibernation. We had no business premises, but our house was quite roomy, and a trestle table with a row of starter cups soon occupied our wide hallway. I had mating cages on the back lawn and in a little glass conservatory. Later, we installed more starter cups in one of the two lounge rooms and laid cheap vinyl on the floor to protect the carpet from spilt sugar-water and pollen.

Figure 5.2 Tables of starter cups in our lounge. **Each table top holds three rows of 22 queens. (1991)**

Keeping bumblebees did not fit into any prescribed business activity. While they were a type of bee and could sting, I considered them to be a totally different activity from keeping honeybees, as it all happened indoors or in cages. But the city council needed to issue a Resource Consent, and we duly consulted all the neighbours in case anyone objected. They were all happy with our keeping bumblebees, but one elderly neighbour had recently died, and his home was being sold. The lawyer for the estate suggested to the man's family that a pending application for consent to keep bumblebees may impede selling the house, and he submitted an objection to the council. He, Rachel and I were summoned to a Council hearing where he made the interesting legal point that a 'fear' of something, even if materially quite safe, can be a valid argument against that activity. When I responded that none of the neighbours I had spoken to, including his client (the deceased man's family), had expressed any fear, he asked that my comment be struck from the record. One councillor agreed with the 'fear' argument, having been stung by a swarm of honeybees himself, but the majority were in our favour, and we received the Consent. As the business expanded, we went through the same process several times again. No one ever objected, and the Council's main concern was always over car parking and what waste materials might go down the drain.

Later, the failure of regulators to understand this new industry became a problem for us when the government Accident Compensation Commission (ACC) re-classified us as 'beekeeping' and charged us thousands of dollars for underpaid levies. The workplace risk profile for conventional beekeeping is heavily affected by injuries from vehicles, burns (from hive smokers) and moving heavy items. Even the risk of stings is minor with bumblebee keeping because they are mostly confined, and you do not work with them flying around you. We did successfully regain our lower-risk classification.

First sales

The founding customer was Tony van Ryan from the Waikato area, who knew that bumblebees were sought after in his native Netherlands. He was pleased with the results, word got around, and by the end of 1991, an enthusiastic market was emerging.

Bumblebees were a highly cost-effective product for greenhouse tomato growers. For a similar cost to employing people to hand-pollinate with electric vibrators, the grower was likely to get at least a 10% increase in production, with more uniform quality. To service 1600 square metres of tomatoes, you needed one $200 hive, replaced six times over the growing season, thus costing $1200. For those who had been paying for the labour of vibrator pollination, there was minimal increased cost. I estimated a 10% increase in yield would net a $16,000 benefit or a thirteen-fold return on the price of the hives. It was economic madness *not* to use bumblebees. One grower who had routinely been using hormone setting found that when he filled his standard-sized cartons with bumblebee-pollinated tomatoes to the same level as before, the carton was heavier. It was due to the naturally-pollinated fruit having more dissolved solids. They were literally tastier and sweeter than the hormone-set ones. The early adopters gained a profit advantage over the non-users, but as with most such technologies, it eventually became standard practice and margins naturally adjusted. Then, growers were at a disadvantage if they *did not* use bumblebees.

Growing pains

Demand for our bumblebees expanded so fast that I used to wonder whether we were living in a fairy tale or a nightmare. The spare living room provided enough space for about 220 queens in their starter cups (on the three tables in Figure 5.2). But after the first workers emerged, the colonies needed larger containers, which would demand more space. So we bought a 6 x 3 m (20 x 10 ft) 'Portacom' unit for our back lawn. It needed to be craned over the top of our house, and our need to house the growing colonies was so urgent that the electricity was connected as soon as it hit the ground. By evening there were rows of hives in there.

Figure 5.3 'Portacom' building craned over our house.

The Portacom expansion was sufficient for a few months, but as had happened in Europe, the New Zealand demand for bumblebees for greenhouse tomato pollination mushroomed. We needed to get serious commercial premises, so we took a two-year lease on a commercial building of around 250 sq m (2700 sq ft). These expansions always happened 'on the run', with growing colonies needing to be moved to the new premises. I remember renting a van to hold the tables of starter cups (the same ones shown in Figure 5.2) and carefully driving them across town amid the agitated buzzes of the two hundred or so queens complaining of the shaking of their starter cups. The Portacom stayed on our home property, and reverted to a handy extra bedroom. Inside the leased building, we built ply-clad rooms with fibreglass insulation for the colony expansion hives.

After just one year, the leased building was too small *again*. We purchased a disused manufacturing butchers' premises of 880 sq m (9500 sq ft. Figure 5.4). The faint lettering on the wall in front of the red car said: 'Watson and Lange – the Thrifty Housewife's Butcher'. The place was built for security (at least from the street) with a 5 m (16 ft) high concrete wall around much of the section, with two industrial-grade roller doors. But there was access from behind, and during the years the building had been vacant, it had been heavily vandalised: all of the copper piping from the refrigeration systems was gone, there was graffiti on the walls, and the separate office block was missing its sliding door and a big window, through which phoenix-palm fronds extended into the middle of the room. There were remnants of kids' fires on the floor. The attraction was that the main building contained six large insulated rooms with thick cork walls – originally for chilling meat carcasses – that we could convert to warmed, controlled-climate rooms.

Figure 5.4 The third location – a disused butchers' building.

Inside the courtyard were two old Morris delivery vans, still with air in the tyres but showing registration labels from twelve years earlier. My fourteen-year-old son and some older mates managed to get one going. It would have been a waste of time to get it to the official roadworthiness standard, so I used to tow it to a nearby off-road path along a river bank, and they had a lot of fun. The clutch wore out eventually, so, inspired by watching the servicing techniques of a Lada in a Russian rally, they tipped the van on its side to access the underside and replaced the clutch with one from the other van in the yard.

Employment

By the time we moved into the big building, I was employing five full-time staff plus several casual workers – mostly high school students (my son and his friends) after school. While the more interesting technical developments concerned housing and managing the bumblebees, there was much work in food preparation, cleaning and assembling hives and starter cups. Tasks were varied, and there was generally an enthusiastic hum in the recognition that we were in a pioneering industry.

Production methods

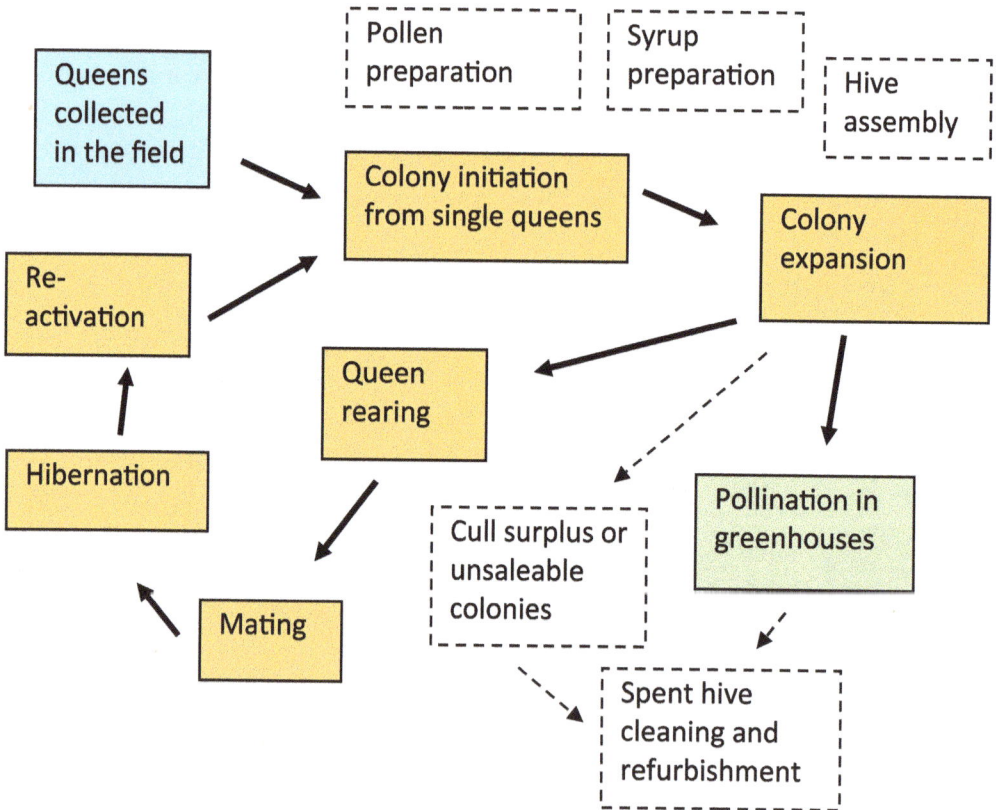

Figure 5.5 Flow diagram of the production phases. **Solid borders and arrows indicate stages of the bees' life cycle. Dashed lines indicate other production tasks. Orange fill indicates the bumblebees' life cycle in the production unit.**

The building had six insulated rooms of about twenty-five square metres each (250 sq ft). One was occupied by up to a thousand starter cups, two or three housed a few hundred expansion hives, and one contained a few tens of colonies for queen production.

Queen collecting

Field-collected queens were our main rearing stock during the early years. The collecting system is described in Chapter 4 (page 76). Local part-time contractors were paid about $1 each for queens delivered to us. As previously mentioned, we instructed that queens with pollen loads were to be released immediately, as they would already have initiated colonies in the wild. Typical sources were roadside tree lucerne in winter and early spring, rhododendron gardens and barberry hedges in later spring, red clover in summer and strawberry tree (*Arbutus unedo*) in autumn.

The volume of bumblebee queens collected from the field internationally resulted in some adverse public reactions, especially in Europe, where they are native species.[7] Here in New Zealand, where they are exotic, there was little reaction. We could only collect queens from accessible locations such as roadsides or public gardens, which was a tiny proportion of the total habitat, so it seems logical to assume we were only removing a tiny proportion of the queen population. I have collected queen bumblebees in public places around Toronto and various parts of New Zealand, and I admit to being self-conscious. I think everybody passing wonders what I'm doing, but few ask. Those that do are usually cheerful about it, but the explanations can be time-consuming!

Colony initiation

Queens were confined in starter cups, as described in Chapter 4. Sixty-six cups fitted on each three-row table, which was on castors so that tables could be pushed together, leaving a temporary working gap. The room was warmed to around 25°C (77°F). The humidity was raised with a steam unit, but we did not consider this critical because the small air volume in the brood compartment of the cup was expected to have its own microclimate, affected by moisture from the food and brood.

Most queens laid eggs within a week of confinement, and if they failed to do so within two weeks, they were usually liberated. Workers emerged three weeks after egg laying, whereupon the incipient colonies were moved to larger containers.

Colony expansion

This is where bumblebee production for pollination in greenhouses diverged significantly from the system I envisaged for kiwifruit pollination described in the previous chapter. Colonies for pollinating kiwifruit, during a brief period in late spring, could live outdoors and gain strength on natural forage, but for year-round tomato pollination, this was impossible even in the mild climate of the warmer parts of New Zealand. So the colonies needed to be kept indoors and fed pollen and sugar.

Although I had designed the starter cup system with an awareness that heating whole rooms to 30°C (86°F) was uncomfortable and uneconomical, the only equivalent local-heating system for the expansion phase would have been something like the heated observation hives (see Figure 2.8). But they were much too elaborate to produce in large numbers. The 'expansion hives' needed to be low-cost, lightweight, well-ventilated, faecal absorbent, and lend themselves to supplying sugar-water and pollen. And we wanted lots of them quickly. We initially chose mass-market plastic terrariums. They were cast in clear styrene and had clip-on slotted lids with hand-sized hinged openings. Sawdust, pumice or shredded paper provided an absorbent bed. In later years we imported Koppert's custom-designed injection-moulded plastic containers with built-in grilles for faecal dispersal and a wick-based sugar-water dispenser.

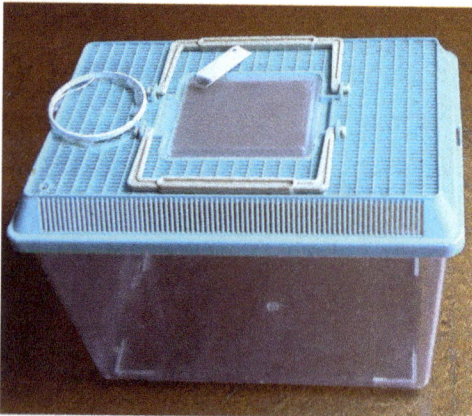

Figure 5.6 Terrariums used by Zonda for exansion hives, and initially as greenhouse hives. **See also Figure 7.1.**

Left.   Single unit. Plastic ring added to position the feeder jar. A lever was added to hold the trapdoor closed (we removed the clip mechanism as the bump on opening disturbed the bees).

Right.   Three units on a shelf with shredded paper base.

The expansion hives were arranged on multi-level shelves in the cork-insulated (previously chiller) rooms which were heated to around 30°C (86°F) with domestic fan heaters and humidified to around 60% RH with steam units. Each unit was an open-topped stainless steel dish plumbed to the cold water supply with a float valve (like a toilet cistern). A domestic kettle element was installed in the base of the dish to boil the water. Full-power boiling was too vigorous and splashy, so the element was 'turned down' with a series-wired Simmerstat. The unit was switched with a humidistat. Staff varied in their enthusiasm for working in these tropical-climate rooms, although they became more popular in winter, as the main building was warehouse-like and unheated.

The warm rooms and the pollen in the hives attracted the stored product pest moth *Plodia interpunctella*, which often bred in the hives, mainly consuming pollen and making unsightly webbing. I experimented with breeding up small parasitoids, which we also saw in the rooms. But a more effective remedy, *when the rooms were empty of bees*, was to spray the walls and shelves with the residual insecticide permethrin. It might sound shocking to use an insecticide around bees, but I was assured that permethrin has zero vapour pressure (i.e. does not evaporate into the air), so the bees were quite safe if returned after the rooms were dry and had been ventilated. Later we used the bacterial insecticide *Bacillus thuringiensis*, which is specific to the caterpillars of moths and butterflies, and is harmless to other insects. We actually mixed it with the pollen that the bees ate. We did not see any *Mellitobia*, a tiny parasitoid of bumblebees, although it does occur in New Zealand and has occasionally been seen in bumblebee colonies outdoors. Nor did we have any trouble from wax moths.

Dispatch to greenhouses.

Overnight parcel couriers delivered the bumblebees. The hives were well secured, and the company was keen for the business, although at one stage, when they centralised their communications in Auckland, and I enquired about a pickup, I was told, 'No, we don't carry live bees.' Well, they did. It cost us about $5 to have a $200 hive shipped anywhere in the country (this was the early 1990s), although customers in rural areas sometimes needed to collect their hives from the nearest town.

On a busy day, we sometimes had thirty hives of bumblebees dispatched to greenhouses ranging from Auckland (450 km north) to Christchurch (730 km south, on another island). They were collected by the local courier around 5 pm and reached their destination area the following day. We discouraged late orders, but if a customer felt it was urgent, we could generally get a hive ready with an hour's notice. Air New Zealand had an over-the-counter parcel service, and on at least one occasion, I took a hive to the local airport on a Sunday to go to Christchurch on the next flight.

Initially, the colonies were sent to the greenhouses still in their terrarium expansion hives. Extra shredded paper or cotton insulation was placed over the comb, and they were shipped in a cardboard box that incorporated a sun-shade flap that could be erected A-frame style. Sugar-water continued to be supplied from an inverted jar gravity feeder placed on the roof. Later we developed a more elaborate greenhouse hive that incorporated a life supply of sugar-water and other features. I describe it in the next chapter. And later in the 1990s when we had adopted the Koppert expansion hives, these were clipped into a standard imported Natupol hive as used by Koppert and its subsidiaries worldwide.

Queen rearing, mating and hibernation

In the production system for kiwifruit pollination, I had been anticipating obtaining new queens from hives retrieved from the orchards after the brief pollination period, but this was not practical in greenhouses because the hives were used long-term and the colonies lived out their natural lives there. So we kept back a selection of good-quality colonies, moved them to larger containers and kept them in the warm rearing room until they changed to queen production. New queens were picked out when a few days old and were kept in isolated groups with plenty of pollen and sugar-water.

Males were also removed from the queen-rearing colonies but were segregated to ensure that queens were never mated with a male from their own colony. It is well-established that brother-sister mating in bumblebees results in around half of the next generation colonies producing a mixture of males and workers from the beginning of the colony cycle when they 'should' be producing only workers.[8] These are diploid males: see Appendix 2. We placed up to one hundred queens plus two hundred males in walk-in cages in a well-lit room, and inspected them every fifteen minutes, as copulation usually lasted at least twenty minutes. Copulating couples were picked off the wall and placed in boxes; the males were later removed, and the queens were confined for another week or two with ample food.

We were working on too big a scale to allow queens to dig in to soil trays (as described on page 87), so they were either placed communally in shallow boxes lined with damp peat or placed individually in matchboxes. In the early days, I believe we ramped the temperature down in three stages: 15 – 12 – 5°C (approx 60 – 50 – 40°F), but later we excluded the 15°C stage. We bought a walk-in cool room for the 5° stage.

Re-awakening

Ideally the queens would be allowed to emerge from hibernation in their own time with gradually rising temperatures, but this was impractical at scale. After a few weeks or months in the chiller, they were taken out and placed in groups in small flight cages (half to one cubic metre (15 – 30 cu ft)) with pollen and sugar-water. Within a week or two, some would become 'broody' and deposit wax on the cage frame. At this point, they needed to be removed to a starter cup; otherwise they would fight with one another.

Later, we would remove the queens from the flight cages after a shorter time (before getting broody), anaesthetise them with CO_2 for a fixed time (a few minutes from memory), and then place them in starter cups. This method was described by Peter-Frank Röseler already in 1985 and is often regarded as the breakthrough that allowed year-round bumblebee rearing, as it was sometimes used to make queens broody soon after mating without any intervening hibernation.[9]

This chapter was intended to outline the production steps we used at Zonda Bees. In the next chapter, I provide more details of the production process, the larger picture of managing the business, and my eventual departure.

Chapter 6

Industry development and personal challenges

Industry development and personal challenges

This chapter provides more details about how we kept the bumblebees. As the industry matured, competition arose, and we diversified into producing other insects and mites for greenhouse pest control. As the company evolved, my place in it became more tenuous, and I describe the restructuring and my personal change of direction.

Feeding the bumblebees

Whereas honeybees are kept on a 'free range' basis, and the apiarist usually has little involvement in their feeding other than some winter or spring sugar supplementation, bumblebees are reared indoors and must be fed pollen and nectar (or its equivalent). All types of bees need an equivalent of nectar as an energy source – almost any sugary liquid will do, and they need pollen for all their other nutrients, especially proteins and fats.

Pollen

So far, there has yet to be a complete substitute for pollen to feed bumblebees. The bumblebee industry depends on the honeybee industry for the many tonnes that the bumblebee rearing industry uses annually. Beekeepers place 'pollen traps' at hive entrances, so returning foragers must squeeze through mesh, which rubs off most pollen pellets from their hind legs. The traps must be emptied every few days, and the product is stored frozen. During the 1990s Zonda Bees paid $10/kg (about $4/lb) and consumed about two tonnes each year. In 2020 the New Zealand price was around $40/kg. Most published bumblebee rearing methods recommend adding moisture (sugar-water or diluted honey), so the granules can be moulded into shapes. (Each granule is the pollen load from one leg of a honeybee.) Initially, when feeding colonies in terrariums, I dumped loose pollen granules into a shallow dish or sometimes placed granules on top of the lid. Some pollen would fall through the slits and be eaten promptly, and other workers would bite at the granules from below until most of them were consumed. This was potentially a good way to avoid pollen wastage because you could avoid adding extra pollen if there were uneaten granules still on the slots. However, the slots became clogged with old pollen over time, and the portion that fell through could easily land on faecal matter and become contaminated.

I really wanted a disposable casing into which pollen could be compressed (thus keeping the air out) and that the bees could progressively feed from without walking all over it with their dirty feet. In Chris Plowright's lab, we had extruded rods of moistened pollen from an open-ended syringe (or a substitute made from a pair of cork-hole borers – the inner one cork-plugged). I found that fresh-frozen pollen could be extruded to a solid shape with little or no moistening if compressed with sufficient pressure. And if it was then coated with wax, it stayed fresh. So we used a hand-wound tool rather like a drill press – it was built by a local engineering company, Haden and Custance, by repurposing the steering rack of a Morris Minor car. Turning the handle forced a plunger down inside a steel barrel. Pollen granules became compacted in the barrel, which could be opened at the bottom so that a further depression of the plunger would eject a solid 'stick' of pollen about 16 x 60 mm (5/8 x 2½ in). The pollen sticks were fairly fragile, but they could be picked up by impaling them on a 'comb' of four fine pins protruding from a piece of wood. The impaled pollen stick would be immersed in molten wax and then dunked in cold water to set the wax so the sticks could be handled and stacked without sticking together. The same engineering company later built us a larger-scale version that used a two-tonne hydraulic ram to extrude about two litres of pollen granules into a continuous series of sticks. The bees usually started feeding at the holes left by the withdrawn pins and enlarged them as required, eating into the pollen and leaving the wax shell, which usually collapsed when empty. The larger machine built in the early 1990s is still in use thirty years later.

We wanted the wax to be malleable by the bees so they could enlarge the feeding holes, so we obtained the soft type used for coating cheese. Sometimes it was blended with some beeswax or paraffin wax. For several years the wax melting pot was an electric frying pan that Rachel and I had been given as an engagement present but had been superseded in our kitchen.

Sugar-water

Sugar, in its various forms, is the energy source for bees. Despite the complexity of honey, and humans' delight in it, I did not believe honey was a nutritional necessity for our bumblebees – I assumed they would receive all their nutrients from pollen. It was just a matter of providing sugar calories as practically as possible. I found it interesting that our partners Koppert in the Netherlands, buying tonnes of sugar, were paying more per kilogram than we were paying by the single sack. This was a result of the agricultural subsidies in the EU. We visited a sugar beet processing factory in the Netherlands. It only operated during the beet harvesting season. Since the late 1980s New Zealand's subsidies for farming have been a small fraction of those in the EU and North America. As a small, isolated country, we have elected to participate fully in, and expose ourselves to, the global economy.[1]

As described in Chapter 3, sugar-water was dispensed via wicks. A problem with sucrose (white sugar) is that it crystallises readily, which tends to block the holes in liquid feeders or the exposed ends of wicks. Initially, at Zonda, we purchased 200 litre (44 imp, 55 US gal) drums of high-fructose syrup, a mixture of glucose and fructose that does not readily crystallise. It is actually close to the composition of many honeys. We eventually stopped using it when we learned that our supplies had relatively high levels of HMF (hydroxy methyl furfuraldehyde), which is a by-product of the formation of the syrup from cereal starch, and is toxic to bees. HMF also forms when honey is over-heated. So, we moved to a mixture of white sugar and glucose (dextrose), purchased dry in sacks and dissolved in hot water in a 200 litre electric stirrer. Caster sugar was almost the same price as regular sugar and the finer crystals dissolved more easily. Using 20% glucose prevented crystallisation.

Those who shun additive-rich soft drinks (as I do) may be taken aback that we added similar ingredients to the sugar-water. Fifty per cent sugar-water does not undergo yeast fermentation as readily as diluted honey, but moulds form in it, showing as black threads or gelatinous blobs. To inhibit this, we followed Koppert's recipe, adding 0.05% potassium sorbate (E202). Sorbic acid occurs naturally in many fruits, where it also seems to play a role in protection from rot microbes. I tried feeding bumblebees with massively increased sorbate concentrations but could find no harmful effects. Food-grade colouring was also added. This was to more easily see the level of the otherwise clear sugar-water in the feeder jars. And by alternating colours for each sugar-water batch, we could see how old it was.

Pure sugar solution has virtually no scent, and honey was dabbed on the top end of the wicks to help queens initially find it. We also added scent to the sugar-water so that when workers were present, they would detect it in the nectar pots and more easily locate the feeder. This was hardly necessary in the rearing unit, but I had thought there might be a case for providing sugar feeders in the greenhouses outside the hives, and the scent would make it easier to find. I used clove oil for a while, which has such an intense odour that only tiny amounts are needed (dissolved in alcohol first so it would blend with the aqueous syrup). I reasoned that clove would not be a naturally attractive scent – it would only have been learned by 'my' bees – and so should not attract honeybees or wild bumblebees. We never did use external feeders and changed from clove to artificial raspberry scent (available as a food flavouring).

The Zonda greenhouse hive

Our first major equipment change was the replacement of the terrariums as the greenhouse hives. Their biggest drawback was that the sugar-water (essential due to tomato flowers lacking nectar) could only be supplied via inverted jars. The customer needed to check and refill these 200 ml (7 oz) jars at least every week, which was a messy job, easy to forget, and sometimes the sugar-water leaked into the hives. Initially, we gave instructions for sugar-water to be made up from white sugar and water, but by customer demand, we were soon selling pre-mixed syrup – the same as we used in the rearing. Some customers also wanted one-way traps on the hives so the foragers could be locked inside and the hives taken out of the greenhouse during pesticide spraying. We did sell a wedge-shaped one-way device that fitted under the trapdoor-style lid of the terrariums, but it was cumbersome.

So, we designed a greenhouse hive that included space for a one-litre (34 oz) sugar-water feeder, a lifetime supply, with no intervention from the customer after opening it on arrival. The one-way entrance system was elaborated to a trapping vestibule on the principle described in Chapter 3. A problem with trapping vestibules for colonies foraging on nectar-less plants such as tomatoes (and kiwifruit) is that the foragers tend to go out with just enough 'fuel' (nectar) in their crop to last the foray. They can starve if they are too long delayed returning to the colony to top up. This was solved by providing a tunnel between the trapping vestibule and the feeding chamber, which was only open when the trap was 'set'. Whereas when the trapping system was not set, the feeding chamber could only be accessed by first going into the brood chamber. This was intended to deter honeybees or other robbers from stealing the sugar-water. It was expected that the bees on the comb would be sufficiently defensive.

The lid and base were vacuum formed separately (by Nolan Plastics in Palmerston North), and the lid flange was heat-folded over, so the lid slid onto the base. I made the initial forms carved from MDF board and aluminium. Michael Nolan took plastic casts off those forms, which I then used as moulds to cast the parts in aluminium-filled epoxy, and then the parts were glued to a 6 mm (1/4 in) aluminium plate to form the mass production moulds. The raw hive shells were only a few dollars each, but there was quite a lot of finishing work for us in trimming, punching and heat-folding.

The customer could trap the foragers from each hive and count how many were working the crop. This was the ultimate in potentially self-damaging transparency: the customers could directly measure how well the hives were doing. It did require the operator to remember to return to un-set the trap, or the colony would suffer.

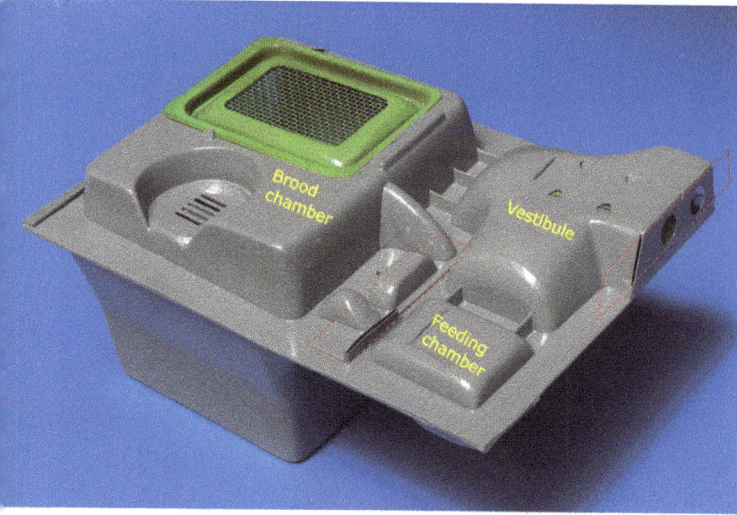

Figure 6.1 Vacuum-formed greenhouse hive. These were made of clear PETG, but this one has been painted to better show the external shape. Red dotted lines represent sliding components. The one near the centre was underneath the surface and was moved via a vertical tab that protruded through the slit.

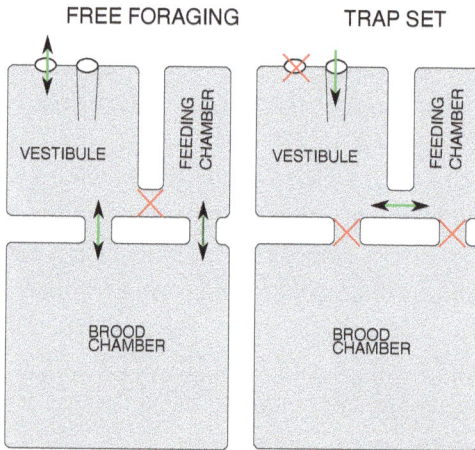

Figure 6.2 Schematic showing the function of the forager-trapping system.

Figure 6.3 Colony in a greenhouse hive. The blue and black circle is the top of the sugar feeder, which was a one-litre rectangular bottle with an access tube of rough porous 'weeper' hose. The white rectangle is the slider to control bee paths for forager trapping.

Previously, we just needed to add some insulating nest material to a terrarium before shipping the colony to a greenhouse. But now we had to transfer the bees and their comb from a terrarium to the new greenhouse hive. Transferring a colony with a hundred defensive workers from one container to another is not easy or safe without some special tricks. Marit was our bold master of transferring colonies. Despite the vandalism of the main refrigeration equipment before we bought the building, an operating walk-in cold room survived. It had a strong fan in the overhead chiller unit, and we made a duct from polythene film, which directed the cold blast downward onto the work bench through a 100 mm (4 in) wide opening. When a terrarium was placed under the cold blast, even the angriest workers could not fly, and the whole comb plus bees could be scooped up and placed in the greenhouse hive. And then, the lid was slid on. The cold blast was not always 100% effective, and Marit did receive the occasional sting. Bumblebee stings were a potential hazard in our operation, and although the handling systems minimised them, we held adrenaline and syringes on site and ran training sessions on treating a person for anaphylactic shock. We never did have a case of anaphylaxis, but one person did develop a sufficiently severe reaction to require medical attention.

Later during the 1990s Zonda changed to Koppert's equipment for both the expansion and greenhouse hives.

Managing production volumes

During the first two years, we just expanded production as fast as we could – chased by the demand from the greenhouse tomato growers – but limited by the number of queens we could collect, facilities we could set up, and sometimes by cash flow. Just once, I paid the wages from my personal Visa account. Queen availability and quality were less in late summer/autumn, and during our second year, we over-committed ourselves. After customers had been steadily buying hives for months, they were not amused to be told we were running short. There were many difficult conversations and at least one abusive one, as we rationed hives, giving priority to our oldest customers.

Even after supply and demand levelled out, neither parameter was certain. The percentage yield of strong colonies fluctuated over the year, there was an eight-week delay between installing a queen and selling the colony, and the sales volume was never smooth. So, to avoid having erratic shortages, we needed to produce a surplus number of colonies. At least this gave us a buffer for replacements under guarantee (see next section).

Bumblebee colonies do not have a fixed size – they become more populous over a period of weeks (or occasionally months) and then decline (see foraging strength over time in Figure 3.18). *If* a tomato greenhouse was an optimal environment for colony growth and *if* we knew that a colony would not change to early male production for other reasons, it would benefit both the bumblebee producer and the tomato grower to send the colonies to the greenhouses while they were still small. They would have been less expensive to rear in terms of labour and pollen consumption and would have maximum future growth and pollination potential. One caveat to this was that until the late 1990s most greenhouses in New Zealand were small – less than 1500 sq m (16,000 sq ft)– and only needed one bumblebee hive at a time. Consequently, that hive needed to be already strong on arrival rather than relying on it gaining strength over time. So, in the early years at Zonda Bees, there was an expectation that a colony would be strong enough to do useful tomato pollination as soon as it was dispatched. But then the question was, 'For how long would it remain at a useful strength?' Our full-priced, highest quality colonies were selected as showing no signs of pending maturity and decline. They needed to have no queen brood, no emerged males, and not have egg-laying workers, which are associated with the breakdown of the queen's dominance and is usually followed by a change from rearing workers to rearing males and queens. The presence of egg-laying workers is relatively easy to recognise: instead of all the egg cells being in single-level neat rows or clusters, there will be untidy stacks of egg cells, and often there will be more than one egg cell open. The latter indicates that an additional bee other than the queen is constructing egg cells, or that some egg cells are being broken open and the eggs cannibalised. This mutual egg-eating between the queen and broody workers is typical in mature bumblebee colonies. In the absence of these signs, we could infer that the unhatched cocoons contained worker pupae, so the colony would likely continue to gain foraging strength and remain strong for several weeks.

Zonda's hive guarantee

Our expectation to the customer was that a colony would provide around six weeks of useful pollination. We guaranteed them for four weeks: if a colony was too weak before four weeks, we would supply a replacement, discounted on a *pro-rata* basis for the proportion of the four weeks lost. So if a colony failed after only three weeks, a new hive would be sent with a 25% discount.

Competition

Some in the greenhouse industry were very aware of Zonda's monopoly advantage, and although the tomato growers could not question the economic benefit of using our bumblebees, everyone had an incentive to reduce costs.

Around 1995, New Zealand interests collaborated with Koppert's main European competitor, Biobest from Belgium, to set up a second bumblebee production company in Hastings. We had reason to believe they would also be producing the same biological control products that had been Koppert's mainstay in the Netherlands and which hitherto had been produced in New Zealand by an Auckland company. We felt that a new company producing both bumblebees and 'beneficials' used to control greenhouse pests would have a real advantage, so we decided to also diversify into those products. The new company was partway through building new purpose-designed facilities when the project was cancelled. The building was never finished. But New Zealand investors took over, including the existing producer of biocontrol products, and a bumblebee competitor duly emerged – 'Biobees'. Within a year, bumblebee prices tumbled from $235 to $150 for a full-strength hive. We scrambled to start producing the 'biologicals'.

Biological control

Two distinct groups of food producers are reducing the use of toxic pesticides. One is the 'organic' movement, and the other consists of 'conventional' growers who use Integrated Pest Management (IPM). 'Organic' production avoids synthetic pesticides and fertilisers (along with other environmental protections), but the distinction between 'synthetic' and 'natural' does not inevitably align with dangerous versus safe. For example, natural pyrethrum insecticide (extracted from pyrethrum flowers) is toxic to all insects, including bees. In contrast, some of the modern synthetic insecticides are specific to only a narrow range of insects and do not kill bees. Copper used as an 'organic' fungicide is more toxic to humans and wildlife than some synthetic fungicides.[2] IPM is usually a mixed strategy, using conventional fertilisers and minimising pesticides, with the motive of minimising the loss of beneficial insects and minimising food residues.

Before the arrival of bumblebees, pesticides were widely used on greenhouse vegetables, with the main precaution being to observe the correct 'withholding period' between spraying and marketing to reduce residues. But with live bees in the greenhouses the whole season, there was an obvious constraint to using insecticides.[3] Whitefly (*Trialeurodes vaporariorum*) were the major pest on tomatoes, spider-mites on cucumbers and thrips on capsicums. Bumblebees were not needed on cucumbers and seldom used on capsicums, but growers were happy to minimise their exposure to toxic sprays when alternatives became available.

Thrip predator

Thrip predators (*Amblyseius cucumeris*) were a mite that could be raised by allowing them to feed on another mite, the soft-bodied *Tyrophagus putrescentiae* that in turn, fed on yeast. Predator-free yeast mites were reared in a small climate-controlled room in bins of moistened bran and (inactivated) yeast. When the yeast mite population density was high enough, the bin was moved to a separate room and infused with predator mites which duly multiplied therein. When most of the yeast mites were eaten, the bran mix containing the predators was packaged up and sent to capsicum growers. Our major customer wanted the mites supplied in tiny sachets that could be evenly distributed around the crop, so we purchased a tea-bag-making machine. Another time his need for large numbers was extreme, so we sent the mite-bran mixture in twenty-litre (4 gal) buckets on a pallet.

Whitefly parasitoid

'Parasitoid' is the technically correct name for a species that lays eggs in or on another species (the 'host'), and the emergent larvae slowly consume the host from the inside without killing it until the parasitoid completes its growth. (These may be loosely called 'parasites', but that is incorrect as a parasite normally does not kill its host.) The parasitoid passes its pupa stage in or on the dead shell of the host, and a winged adult ultimately emerges. The parasitoid of whitefly was the tiny *Encarsia formosa* (Figure 6.4).

As with the thrip predator mite, there needed to be food for the parasitoids. In this case, we used whitefly, which we raised on tobacco plants. Tobacco had the advantage of being a favoured plant for whitefly, and the big leaves were practical for harvesting the *Encarsia*. So we needed a greenhouse to grow the plants. We leased a back section a few kilometres from the bumblebee production unit and built a 600 sq m (6500 sq ft) double-skinned plastic greenhouse. Inside the greenhouse, we built a series of walk-in, insect-proof compartments (walled with 1 mm (1/25 in) mesh). The key was to maintain a pure culture of whitefly – totally isolated from the parasitoids or any other predators, then to harvest enough whitefly to massively and synchronously infest tobacco plants in a new cage. The progeny of this mass infestation remained parasitoid-free until half-grown, and then enough *Encarsia* were introduced to the cage to lay eggs in all the juvenile whitefly. When the thousands of *Encarsia* had reached the cocoon-stage, the leaves were removed, the cocoons blown off with water jets, dried and stuck to small cards for dispatch to greenhouses.

Figure 6.4 *Encarsia* next to a pinhead.

Encarsia production, more than any of the other rearing systems, opened my eyes to the massive productivity of 'farming' systems compared to a natural environment. By exploiting the gross reproductive capacity of an insect, providing abundant resources for growth, and eliminating enemies, vast numbers of individuals could be produced on a small footprint of space and resources. In that one modest greenhouse, we could rear enough *Encarsia* to potentially control whitefly in over a hundred hectares of tomato greenhouses. *Encarsia* became a major product, and at one cent each, the annual revenue from them by the late 1990s was similar to the income from bumblebees.

Spider-mite predator

Two-spotted spider-mite *Tetranychus urticae* infests many plants, lurking under fine webbing, feeding on and damaging leaf tissue. They are eaten by the predatory mite *Phytoseiulus persimilis*. As with the whitefly parasitoid, mass production depends on rearing vast numbers of predator-free two-spotted mites and then using them as prey to rear predator mites. We built a second 600 sq m greenhouse for this. The mites were reared on dwarf bean plants. Separation of the host from the predator was less critical in this system, and although the greenhouse had a central partition, separation was mainly achieved by differing humidity levels.

In all rearing systems with separation between host and predator, there also needed to be strict protocols on staff movements to avoid cross-contamination on clothing. After working in a predator environment, it was normally prohibited to enter a predator-free compartment the same day or without a change of outer clothing.

Restructuring and my exit

During our early expansion phase, we employed Warren Hobson, a Hawkes Bay local who had worked at Koppert in the Netherlands. Warren supervised bumblebee production and visited customers. In 1994 I gave up my title of Managing Director and promoted Warren to General Manager. I spent more time as the 'boffin' and called myself 'Technical Director'. I worked on various biological questions and co-designed and built new equipment.

During the late 1990s there was a cascade of influences that culminated in my departing from Zonda Bees. It was preceded by, but was not necessarily caused by, my marriage to Rachel breaking up. As part of the standard 50/50 division of marital assets, we agreed that I would buy her 25% shareholding. Koppert agreed to purchase 20% shareholding from me, as I was taking a lower profile in the company operations, enabling me to finance the transaction with Rachel. So then the shareholding was divided 70/30 between Koppert and me.

Risk and influence

Despite the notion of 'limited liability' of companies, it is normal for all banks and other creditors to require directors to sign guarantees to cover any liabilities the company might incur. I had long been aware that Rachel's and my guarantees were the only effective ones: the other director being offshore. It was our house that Zonda's bank held a mortgage over, even though it was paid off. So now, I was effectively carrying the total liability for the company while being a minority owner. I pressed to get the banking and guarantees restructured without success, so I began to look for alternative opportunities.

Selling out

Warren had moved on, and Zonda was run by a 'management group' of myself and three other staff. There were some problematic employment decisions, and I felt that carrying my director's guarantee in the face of minority control made my position untenable. I wanted to get out. I hoped Koppert would buy me out, but they wanted New Zealand interests to do so, which looked like we needed to sell the whole company. So I started looking for alternative owners, and we came within reach of the competing company taking over. But a management group member solved the problem by raising funds to buy the share balance herself. So I was out by the end of 2001 and had a refreshing tramping holiday with one of my sons and friends in Nelson Lakes National Park. My question was what to do next.

In preparation for my exit from Zonda Bees, I formed my own company, imaginatively named 'Nelson Pomeroy Limited' (NPL). I wasn't sure what I would do, so I avoided descriptive words like 'Bumblebees' or 'Pollination'. I had in mind developing observation hives of bumblebees for schools or doing contract research for Zonda.

Zonda Resources Ltd moved operations to Auckland, continued for another ten years, and after some difficulties, went into liquidation in 2011. The assets were bought by New Zealand Gourmet, a large horticulture company whose operations included blueberries, capsicums, and tomatoes. It trades as 'Zonda Beneficials', and an ongoing Koppert association is apparent in their Natupol branded greenhouse hives.[4]

Nelson Pomeroy **Bumblebee Keeper**

Chapter 7

After Zonda Bees: travel, school and kiwifruit again

After Zonda Bees: travel, school and kiwifruit again

From my late teens to my end-forties, my career comprised about equal decades as a student, a university academic, and a bumblebee producer. My twenty years since leaving Zonda Bees have been dominated by a decade of high school science teaching. I trust the reader will forgive my digression from the bumblebee theme, although I sometimes had bumblebees in my classroom. Kiwifruit pollination kept on coming to meet me too.

Varroa mite and a kiwifruit interlude

During my final year with Zonda Bees, I also had a separate company (NPL). My first research contract involved kiwifruit again.

In 2000 the *Varroa* mite took hold in New Zealand, and the honeybee industry was reeling. We had seen this happen already in other countries, and the New Zealand pattern was similar.[1] Unprotected colonies were being wiped out, and soon it appeared that the entire feral population had been eliminated, at least in the North Island. Honeybee hive supply and prices were likely to be affected, and the kiwifruit industry was nervous. The bumblebee rearing industry had begun a decade earlier, and a few bumblebee colonies were already being sold for kiwifruit pollination as an adjunct to honeybees. But my earlier trials (Chapter 3) suggested that several tens of bumblebee colonies would be needed per hectare to *replace* honeybees, and this was likely to be too expensive for the kiwifruit growers. However, the 1980s trial was on a small scale, and I was keen to extend it. I obtained a contract from the kiwifruit industry ('Zespri Innovation'), and with one foot still in Zonda, I (NPL) was able to raise colonies for a larger trial. I had over seventy peak-strength bumblebee colonies ready to place in an orchard of the up-and-coming gold variety ('Hort 16A'), but a storm damaged the crop, and so as a fall-back option, I was offered a green ('Hayward') orchard for the trial at Te Puna. Hayward was later-flowering, so I needed to hold the bumblebee colonies for a few extra weeks, during which time they were getting past their peak. As discussed in Chapter 3, this trial used bigger cages, each covering several vines. The results were consistent with the smaller earlier trials and still suggested several tens of bumblebee hives would be needed to fully pollinate a hectare of kiwifruit. At the then price of $150 per hive, the cost was going to be several times greater than honeybees – possible, perhaps, if honeybees became unavailable, but certainly well above grower expectations.

But the trial showed other possible advantages of bumblebees. There was evidence that honeybees were spreading over an extensive area – there were already honeybees on the kiwifruit flowers before the beehives were brought in, and no noticeable increase afterwards. This suggested that honeybees operate on a district-wide basis and that individual orchardists were not necessarily receiving the direct pollination benefit for which they had paid.[2] The large foraging radius was well-known, but my bee counts on flowers added confirmation.[3] Bumblebees, on the other hand, appeared to be confining their kiwifruit foraging to that orchard. As explained in Chapter 3, pollen grains from female kiwifruit flowers were a different shape from grains from male flowers. By microscopic analysis of pollen loads from the legs of honeybees and bumblebees, I could show that bumblebees were crossing between male and female vines more than honeybees. Bumblebees were also flying at cooler temperatures.

But, despite the advantages of bumblebees, the colonies were too small and/or too expensive to compete with honeybees. My recommendation was that further research should be aimed at finding ways to produce bumblebees more cheaply, with larger colonies, or both. There did not appear to be a positive response at the time. But a decade later, bumblebee research was being supported again, and I returned to it in 2014 (see later this chapter).

Overseas travel

On the first anniversary of the 2001 terrorist attack on the World Trade Towers in New York, I flew to Britain to support my 89-year-old father on a visit to his family (and an old girlfriend) in Devon. While he returned separately, I visited some companies in Europe and USA to see if I could engage in consultancy for bumblebee production. I was generously hosted, but I don't think I was ever much of a salesman. At least, an expert as far away as New Zealand was not very interesting to them.

Alex Chu from Massey University had a long association with China and facilitated business collaborations with agencies there. He contacted me with the invitation to meet with Chinese officials who were interested in bumblebee technology. The result was an invitation to visit the Beijing Institute for Agriculture and Forestry as a guest of their government. I had read up on doing business in China and hoped to get a consultancy agreement to collaborate with them as they developed a rearing program.

It was a fascinating week, from the big red sign in Beijing airport welcoming visitors to China and pointing out that you must abide by the country's laws, to the huge murals on high-rise buildings showing how handsome and attractive the police were. I was treated very hospitably and was embarrassed by how free my hosts were to spontaneously buy me gifts during touristic outings. With a couple of young interpreters, I showed more than perhaps was prudent of the technology of bumblebee rearing systems. Although I tried to negotiate a consultancy agreement, I did not know how to work the system and went away chalking it up as a fascinating experience.

One surprise during my visit to China was the suggestion that 'surely the bumblebee production system could be automated'. I responded that I did not think this would be a priority with China's low wage costs. However, I was told that 'the government wants things to be high tech', and that person had a friend in the computer industry who would be travelling some distance to talk to me about it. It came to nothing during the visit, but it made me think after returning home. I had always had a hunch that bumblebee colonies would do better if they were not over-fed – intelligent rationing would also economise on expensive pollen. So it seemed to make sense that if the colonies were fed a measured amount of 'just enough' pollen, it might have several benefits. Pollen consumption is known to be largely by driven by larval demand. Larval quantity can be measured as the area seen from above. With modern optical and computing technology (even in the early 2000s), it ought to have been possible to develop such a system, which might involve hives on a conveyer system passing under scanners. I did discuss it with the local engineering company Haden and Custance (who had built the pollen presses), who specialised in automated materials handling technology. But first, I needed to experiment with differential pollen supplies in relation to larval area to see if my hunch was correct. I got as far as building a thirty-degree incubator, resurrecting some old starter cups, and installing queens. But I was not going to be able to do this on a big enough scale, with the resources at hand, to make much progress. So I resumed a project, begun during the Zonda years, of designing a user-friendly observation hive in which to sell bumblebee colonies to schools.

Bumblebees for schools

Bumblebee colonies, well displayed, are fascinating to watch, much more so than observation hives of honeybees. Due to the small numbers and compact dimensions, there is always some action – the queen laying eggs, foragers returning and unloading their pollen, antagonism between broody workers – and all within a breadth of a few centimetres. I thought that bumblebee colonies would be fascinating additions to school classrooms, and even an attractive alternative to fish tanks, in the waiting rooms of high-end professionals. Zonda had sold a few unheated observation hives like the one in Figure 7.1, but the interest in these hives from schools was minimal, although we did provide a few to kindergartens. However, I thought a more custom-designed hive with internal heating would be a better product, and I spent much time working with electronic heating systems and possible hive designs. Although I spent a lot of time tinkering with electronic heating systems and hive designs, it was not looking fruitful. Meantime I was living off the modest capital I received from the sale of my Zonda shares.

Figure 7.1 Observation hive for schools. The mesh tunnel is connected to a window for the bees to fly freely.

Teaching high school science

By the end of 2002, I was running out of savings, and my new partner Mies was running out of patience. I had applied for a couple of university jobs without success, so I considered high school teaching. Back in 1981, when I had just returned from Canada with a gastrointestinal disorder, unsure of my future and whether perhaps it still might be in bumblebees, I had visited some sort of esoteric healer/fortune-teller. As he dangled a quartz crystal over anatomical diagrams and listened to my woes, he said he didn't think my future lay in a bumblebee business 'whatever that is' but in education, 'although not in a conventional way'. He was wrong on the first point; if you count Zonda Bees as my 'future' – it depends on how far ahead he meant. The 'teaching thing' did lurk in my mind sometimes, so perhaps it was a self-fulfilling prophecy. Anyway, teaching paid a steady wage, and I like telling people how nature worked, so aged fifty, I applied for and was accepted into a one-year postgraduate diploma in teaching, with specialities in general science, biology and physics (with Massey University again, but primarily by distant learning from home).

That summer, waiting for the course to begin, I took a job with a farm irrigation company, and for a while, I thought I would stay in that industry. But during the school holidays, there was a sudden vacancy for a science teacher at the Rudolf Steiner school (I'd left Massey University and moved towns in 1990 so my children could go there). My boys had been through the school and my daughter was still in her senior years. I had been open to teaching at any school, and while I was happy about my children attending the Steiner school, my worldview had become more mainstream and materialistic than the typical teacher there. Still, I got the job on a part-time, one-year contract in parallel with my diploma studies.

There seems to be a view that 'alternative' schools are more liberal than state schools and let students choose their learning. This is quite contrary to the reality in Steiner schools, where the curriculum (until relatively recently) has been fixed, and all students took the same subjects. On the other hand, the lack of external exams when I started there meant teachers could exercise autonomy in the content of their lessons.

A feature of Steiner schools is the 'Main Lesson' (ML) , which occupies two hours every morning on a single topic for a three to four-week stretch. While traditional guidelines were available, I sometimes preferred to invent my own content and practical activities. In the 'Hydraulics and Aerodynamics' ML, I thought flotation could lead to some relatable fun. One exercise involved measuring the volume of big plastic bins to calculate whether they would have enough flotation to support a student in the school swimming pool when formed into a raft. This was duly tested.

I also wanted to measure the *density* of a student. I converted my home 180 litre (40 gal) wheelie bin (thoroughly cleaned) into an 'Archimedes can,' with an overflow pipe plumbed into a hole near the top. The bin was filled up to the overflow hole with water (warmed in colder weather). A student selected the previous day and instructed to bring swimwear, was totally immersed in the bin, and the overflow volume was measured. Then s/he was weighed. Density = Mass / Volume. The usual result was around 0.95 kg/L – slightly less dense than water (1.0). We discussed how the student accordingly did float, and had to be pushed down to achieve total immersion. Anyone who has made a stew will have seen that meat and bones sink, so what are the 'light' parts of a human body? We noted that a breath was taken in before immersion and concluded that air in the lungs accounted for the tendency to float. At that stage, I was a volunteer airborne observer for the Coastguard and explained that drowning victims sank after their lungs filled with water.

Students always like explosions, and sodium is a novel substance whose chemistry is instructive (and closely related to lithium as used in batteries). The element sodium is a metal, almost as soft as cheese, and shiny and silver when freshly cut. It reacts violently with water to form sodium hydroxide (caustic soda or lye). It is so reactive that it does not exist in the pure metallic form in nature – only as salts. For laboratory use, it has to be stored immersed in oil. A finger-nail-sized chip of metallic sodium will merrily fizz around a dish of water, sometimes with an orange flame, and can be demonstrated in a lab. But a chunk the size of a stock cube will explode. I always did this outdoors, using several metres of string to pull the paper-wrapped sodium lump off a platform into a large dish of water while we all stood well back. Nothing noticeable happened for quite a few seconds. Then the explosion sprayed drops of burning molten sodium around. These scattered fragments contained unreacted pure sodium – there was a danger that a bare-footed child could step on one, and the skin moisture would set off the reaction, producing both a physical and chemical burn. So we would douse the surrounding area with lots of water (creating another entertaining bout of flames and pops) to react with all of the sodium and dilute the corrosive product.

Students arriving at my lab would sometimes ask, 'Can we blow something up?' and I would reply, 'Yes, you can' and offer them a party balloon.

Figure 7.2 Sodium explosion. The glowing projectiles are droplets of burning sodium metal.

Bumblebees in my classroom

I did improve on the observation hives, as shown in figure 7.1, but I only made a one-off version that would have been too expensive to produce as a saleable product. It was based on a wooden box with perspex/plexiglass on one side for viewing and a cone-shaped false floor with heating underneath. It even had internal lighting, which used a 'press and hold' switch so it could not be left on and thus confuse the bees. My first teaching laboratory was at ground level, with opening windows, so the bees went out through a tunnel fixed to a hole in a board that the window closed against. The school

later acquired a four-storey office block partly converted to teaching space, and my new laboratory was on the third level. The windows did not open, so I drilled a 12 mm (half inch) hole through the aluminium window frame and ducted a colony of bumblebees out through that.

During the 'Environmental Studies' ML the students made bumblebee domiciles out of pumice concrete cast between concentric sleeves of thick cardboard. These had tunnels made of flexible electrical conduit tubing, with the entrances stabilised by casting them into brick-sized blocks of concrete. What a searching queen saw was the concrete block on the ground surface, with a hole in the side, which was the tunnel entrance. These persisted for many years, partly buried in the school grounds, and were frequently occupied by queens.

On the good days, teaching was my most satisfying career. On the bad days, I bemoaned the combination of my (conflict-avoiding) personality, and my lack of training in classroom management. In 2014 three factors came together: I spent a lot of time with a particularly unmotivated class; some parents were questioning my adherence to the traditional outlook of the Steiner philosophy; and I saw an opportunity to return to bumblebee work. I quit partway through the year in time to fit in a bumblebee season but taught part-time to see some classes through to their end-of-year assessments.

Kiwifruit again: an alternative rearing concept

Bumblebees hit the news more often after 2010. A group at New Zealand Plant and Food Research Limited (PFR), headed by David Pattemore, was working on the pollination of kiwifruit and other crops.[4] I was tempted to get in on the action, and in 2014 I arranged to visit David and expressed my interest in the project. It was the first time we had met. But he had already been using my pumice-concrete recipe for field domiciles, and unbeknownst to me, he had been a reviewer for a book chapter I had recently written on the early bumblebee developments at Massey University.[5] So he knew who I was and invited me on board as a sub-contractor to the project. It was like 'coming home', and I was surprised at how emotionally affected I was. When I returned to our motel, I shed a few tears on Mies's shoulder. It was something about feeling wanted.

Kiwifruit are New Zealand's biggest horticulture export at the time of writing (2023), and in 2020 they matched the value of meat exports. The honeybee industry had settled down after the *Varroa* mite outbreak. The mite was still rampant, but beekeepers had a control regime. It was based on an unusual chemical, a type of synthetic pyrethroid insecticide: the same family of chemicals as is used in fly and cockroach sprays. Most pyrethroids are lethal to all insects, spiders and mites (apart from a few that have evolved resistance), but this one did not kill honeybees.[6] There was indeed some concern that the *Varroa* mite may be evolving resistance, and that other remedies may be needed, but for the time being, *Varroa* was under control in managed beehives. The newer 'Gold' kiwifruit cultivars are generally grown under fine net cover as protection from wind damage, and there has been concern over the use of honeybees working in this environment.[7] As well as the price pressures of *Varroa* control, honeybee availability was potentially compromised by beekeepers wanting to divert their hives to hill-country apiaries where wild manuka shrubs (*Leptospermum scoparium*) yielded nectar for the lucrative manuka honey.

Thus, bumblebees were being more seriously seen as alternative pollinators. I have discussed the complexities of kiwifruit pollination in Chapter 3, and I estimated that several tens of bumblebee colonies per hectare might be needed. That was for the green cultivar. The gold ones have fewer seeds and may not need as many bees. But in any case, David and I agreed that high densities of bumblebee hives would be required to achieve anything more than to add a small supplement to honeybee pollination. Zonda Beneficials and Biobees were already mature companies producing a few thousand bumblebee colonies annually, and some were already sold for kiwifruit pollination. At typical prices of $150 per hive, the use of several tens of hives per hectare was likely to be unacceptably expensive for kiwifruit growers. Hence the programme aimed to find lower-cost ways of providing bumblebees for pollination.

David's team had seen bumblebees in terms of 'ecosystem services' provided by the wild population. Working on the assumption that wild bumblebee populations were likely to be limited by the availability of nesting sites (mostly disused rodent nests in burrows), they placed concrete domiciles in the field as additional nesting sites. Occupation of these artificial nesting sites was unreliable, and they tended to attract more long-tongued *Bombus ruderatus* than the more desirable *B. terrestris*.

Providing extra nesting sites for wild queens would not likely have boosted the wild bumblebee population to an economically useful level. Instead, with my background in artificial rearing, we discussed the idea of taking a more commercially managed approach. Part of the reason for the high price of bumblebees for greenhouse pollination is that tomatoes are grown year-round, and thus need bumblebees in all months. This means the production facilities must manage all phases of the bumblebees' life cycle indoors under controlled conditions. But kiwifruit only requires pollination for about two weeks in late spring. Perhaps we could eliminate the costs of indoor queen breeding and hibernation, as well as the cost of feeding colonies indoors while they grew to a saleable strength for pollination. The idea was to capture queens in the field (as we had at Zonda, especially during the early years), install them in a colony-starting system to lay eggs and raise the first batch of workers, and then transfer these small, low-cost, 'nuclei' colonies to outdoor hives to grow, sustaining themselves on flowers in the local landscape. This was essentially what I envisaged in the 1980s. The flow chart of that proposed production system is shown in Chapter 4 (page 75). The existing companies Biobees and Zonda Beneficials could have raised such nucleus colonies, and I do not know the details of PFR's negotiations with them, but they were not involved in this work. I think it is fair to conclude that they did not see it as in their commercial interests to encourage the production of lower-priced bumblebee colonies, or perhaps there was some scepticism over the quality and uniformity of the colonies from such a low-input production system. Our vision was for a separate bumblebee industry for low-cost colonies for outdoor crops that would be less capital-intensive than the big year-round 'factories'. Perhaps this new industry would be more dispersed around the country so that multiple entities could each collect a few thousand queens and rear their nuclei. As the rearing phase would only occupy a few weeks in early spring, we assumed the equipment would need to function without climate-controlled rooms and could occupy seasonally idle packhouses.

My main role with PFR was to rear the nucleus colonies from single queens and to refine the techniques for doing so. My first contract was to raise fifty nucleus colonies. I had a fridge-sized incubator (I'd made in the aftermath of the China visit) in which I raised some colonies, but I preferred the system with starter cups on a heated metal strip (Figure 4.8). This was all happening at the same time as I was still teaching part-time. My contract started too late in the season for a kiwifruit pollination trial. But as an initial performance test, I transferred my nucleus colonies to half-buried concrete domiciles that PFR had installed around local plum and apple orchards near Hastings. To measure colony growth, I built trapping vestibules – shallow, clear-roofed boxes that fitted beneath the concrete lids, to intercept returning foragers. The concept is outlined in Chapter 3, but to reiterate, it works like this: Foragers must pass through a vestibule on their way in and out of the hive. To sample the number of foragers in the field, the hole between the vestibule and the hive is blocked so that no more can come out, and the two-way hole between the vestibule and the outdoors is also blocked, forcing returning foragers to enter via a one-way tube. Thus, by the time all the foragers have returned (they seldom stay out for as long as an hour), they are confined to the vestibule, which can be photographed for later counting. Figure 3.7 shows the principle, although I only used a single compartment, and replaced the one-way 'out' tube with a plain hole.

Figure 7.3 Trapping vestibule with foraging population intercepted. Contains about sixty foragers. The wooden block on the upper right closes or opens a hole accessing the main hive by pulling the wire from outside. The one-way tube is at the lower centre. The blue cap is for sugar-water to prevent starvation of trapped bees.

To estimate the total productivity of each colony, I retrieved the comb when activity had ceased so that I could count the empty cocoons. As usual, there was a huge variation. The smallest colonies, which never had more than five foragers working, produced less than a hundred bees, many of which were probably males (worker and male cocoons are similar-sized). The range is shown in Figure 7.4. The largest colony had up to sixty foragers working, producing eight hundred workers/males and three hundred queens. One other colony would have been even larger – it had nearly a hundred foragers, but it stayed active so long that I had stopped regular inspections, and a rat destroyed the cocoons before I could count them. The median productivity index was about 350, equating to a maximum forager force of no more than twenty workers.

Figure 7.4 Frequency distribution of sizes attained (Productivity Index) by colonies of *B. terrestris*.

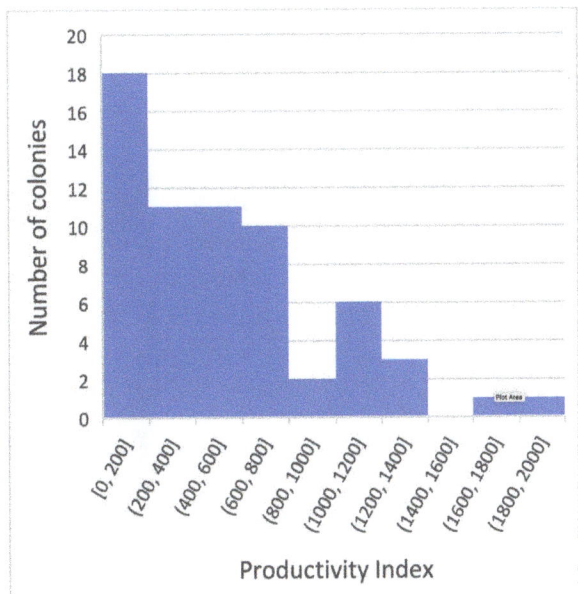

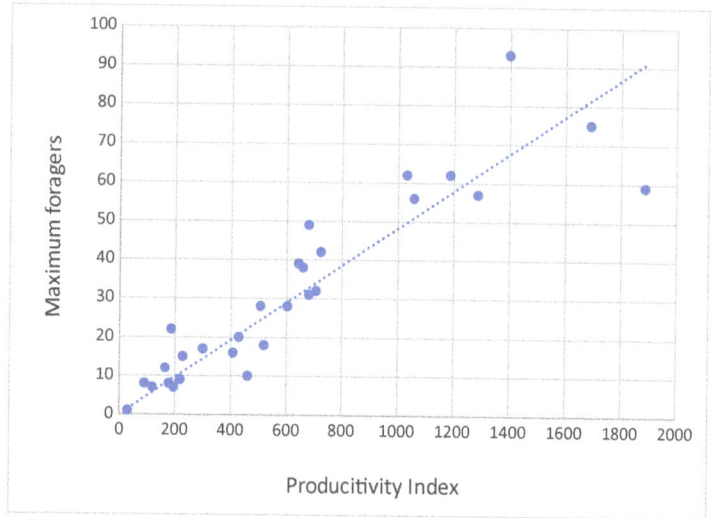

Figure 7.5
Maximum number
of foragers working
from *B. terrestris*
colonies in Hawke's
Bay apple and
plum orchards
in relation to the
Productivity Index.

The following season, PFR planned to run a larger trial in kiwifruit orchards. I was contracted to raise five hundred nucleus colonies of *B. terrestris*, to be couriered to Hamilton for placement outdoors. My project had two aims: the practical production of colonies to be used in the pollination trials, and the assessment of the rearing techniques for commercial scalability. Although the intention was to cast the necessary starter cups by vacuum forming, this was not possible in time, and I reverted to hand-crafting an equivalent device around two-compartment takeaway food containers. It involved a lot of plastic cutting, milling a channel to make access between the two halves, and even heat-and-press forming a plastic insert to provide the bowl-shaped nesting compartment with its knob in the centre and nectar-pot hollow (Figure 7.6).

Figure 7.6. Hand-made
starter 'cup' from a
two-chamber food container.
Resting on the warmed
aluminium strip in a tray.

Figure 7.7 Rearing rack under construction. **The vertical partitions are double layered to allow fan-forced air into them, which exited via small holes near the centre of each tray.**

The scale of rearing I needed to perform over a few weeks was quite similar to what Zonda Bees would have done over an equal time – I just did not have a 'factory' around me for the task, just our home. So it was a good test of mass-rearing without climate-controlled rooms on a small floor footprint. To raise five hundred good colonies, I expected to need at least one thousand queens and an equal number of starter spaces. Thus began the bulk 'rack' system. It was based on 'trays' that held a double row of starter boxes (2 x 10) on electrically heated strips, much like the original patent (page 81). In the original system, as Zonda Bees and Koppert used, the heated strips were permanently wired to movable tables or pull-out shelves, but I needed to make the system more compact. I chose to house the queens in double-row 'trays' inserted end-wise into a 'rack'. The electrically warmed trays were unplugged from the power (12v) and entirely removed for servicing. Figure 7.7 shows a partly-built rack with the tray runners visible that would contain 3 x 6 = 18 trays. So the total unit contained about three hundred and fifty queens in a block about 1.3 m (4.5 ft) cube, occupying less than two square metres (20 sq ft) of floor space. (Additional space was needed to withdraw trays and service them.) Active ventilation was used to disperse heat and humidity build-up. Figure 7.8 shows me with three trays removed for servicing. The rack is visible behind.

It was a wild season: I built three racks in our semi-enclosed home garage and installed thirteen hundred queens. Mies and I caught most of them, but family members in Tauranga also contributed, as did some PFR staff and others locally. I sometimes had two staff assisting with the various construction tasks. But we succeeded – I sent nearly six hundred nucleus colonies to the research group at Hamilton. The colonies were distributed over several kiwifruit orchards, and the PFR team were able to further their research into bumblebee effectiveness on kiwifruit.

Over the next few seasons, I was contracted to deliver more nucleus colonies and refined the starter cup, with PFR providing drafting work and 3D-printing a mould. From their modified mould, I made a set of four with aluminium-filled epoxy to use in the vacuum-forming machine I had purchased for the project. Vacuum forming involves clamping a thin sheet of plastic into the machine, heating it to softness, and vacuum-sucking it onto the heat-resistant mould. The plastic hardens to the mould shape within seconds and is lifted off. I could cast four starter units at a time with a cycle time of around one minute. Over the next two years, I produced over three thousand starter cups for PFR.

Meantime two racks had been shipped to Hamilton, and I built a 'Mark two' version that held five hundred queens on a similar footprint with tidier electronics and insulation. PFR went on to create a modular version, under the name BumbleBox™, which can be scaled to suit the available space and production requirements. At the time of writing, they are actively working with industry partners to commercialise the system for the benefit of the New Zealand horticulture industry. Contact details for anyone wishing to assess the BumbleBox™ as a commercial endeavour are in the end notes.[8]

Figure 7.8. Myself with trays of starter units. The trays were removed from their rack for inspection and feeding. At the working level there are two trays parallel with each other, and a single tray higher up on the processing shelf. In the background is a whole rack with the insulating cover rolled up. The silver rectangles are the ends of trays.

Chapter 8

The environment and bumblebees, rejected or cherished

The environment and bumblebees, rejected or cherished.

So far, I have said little about pollinator conservation, whereas some acquaintances assumed that would be the main subject of this book. Pollinator conservation is well covered in other books written by those more involved in the issue.[1] My impulse generally, if you haven't noticed, has been towards technical development rather than conservation in the classical sense of conserving the status quo or reinstating the past. However, the former can assist the latter, as I will enlarge upon in the final section.

To illustrate the variety of angles to bumblebee abundance and scarcity, I have (in keeping with my style thus far) described some situations in which I have had a personal interest: the option of commercialisation of bumblebees in Australia, the extinction of *B. subterraneus* and the arrival of foreign species in Britain, and the abrupt decline of a group of bumblebees in North America. I conclude with a personal view of options for the future.

Some national examples.

Bombus terrestris in Australia

While neither New Zealand nor Australia has any native bumblebees (New Zealand lacks any native *social* bees or wasps[2]), the fact that they were successfully established in the former country over a hundred years ago makes them an accepted part of the landscape like British willow trees, blackbirds, finches and sparrows. (Not all established species are 'accepted' – there is a constant battle against introduced predators and weeds.) Bumblebees did not become established in Australia following early attempts to introduce them there, probably around the same time as to New Zealand (late 19th, early 20th century).[3] By the late 20th century, the science of ecology was well established, and the conservation of native flora and fauna was given high importance in both countries. Thus, when *B. terrestris* were discovered in the Tasmanian city of Hobart in 1992, they were conspicuously foreign. It was assumed they came from New Zealand, by accident or design, by air or sea transport, the latter most likely as they were first found near the docks and marinas. The 2000 km separation from New Zealand makes self-migration vanishingly improbable. Subsequent genetic analysis confirmed New Zealand as the source.[4] From a conservation point of view, bumblebees were regarded as unwanted and potentially invasive. However, the greenhouse vegetable industry regarded them as a positive opportunity and hoped to set up a commercial pollination supply. Although the bumblebees rapidly spread through most of Tasmania, they remained absent and prohibited on mainland Australia.

We at Zonda Bees also saw this as an opportunity, as rearing expertise would be required if bumblebees were to be commercialised for pollination, but first, we were suspicious of the genetic robustness of the population. If the population had arisen from the accidental arrival of a single fertilised queen, then the population would be inbred. That leads to the phenomenon of 'diploid males' where 50% of what 'should be' workers emerge as males (which do no work). I explain the biology in Appendix 2. So in 1994, I went to Tasmania, and with Roger Buttermore, an entomologist from the Tasmanian Museum and Art Gallery, collected queens in the Botanical Gardens and reared colonies from them to see what the offspring were like. It turned out that half of the colonies produced a mixture of males and workers in their first brood, which is consistent with diploid male production, although we did not verify this with cytology analysis.[5] This suggested that the whole population arose from a single queen and indicated that half of all the wild colonies would be only rearing half of the normal worker populations. The flourishing population seemed remarkable if it was reproducing under this constraint. A more rigorous genetic analysis later confirmed the highly inbred feature of the Tasmanian population.[6]

Zonda Bees had no further technical involvement, although I attended some meetings which discussed the arguments for and against introducing bumblebees to the mainland for commercial pollination. The main environmental risks proposed were competition with native insects, birds and mammals for nectar, and the potential for increased pollination, and hence seed production, from weeds which had hitherto been poorly pollinated by native insects and honeybees.[7] The honeybee industry was also not keen due to the perceived risk of introducing bee diseases.[8] The idea that bumblebees might cause weeds to spread was new to me. But I accepted it to be a valid environmental risk, even though the magnitude had to remain speculative, barring a large-scale and probably irreversible experiment. My environmental sympathies were at odds with at least some of the representatives of the horticulture 'side'.

It soon became apparent that bumblebees would remain prohibited from mainland Australia, and regulations were drafted to prevent them from being commercialised in Tasmania because it would increase the incentive to move them to the mainland. In reaction to pressure from the greenhouse industry, the federal government in 2019 proposed amending the regulations to allow a two-year trial of commercial use of *B. terrestris* in greenhouses within Tasmania.[9] It seems a conflicting regulation classified *B. terrestris* as a 'threatening' species, thereby preventing it from commercial use. As of 2023, there was no commercial exploitation of bumblebees in Tasmania.

The borders between the states of Australia are 'open', and while the movement of certain goods between them may be prohibited, there seem to be no routine checks. This fact, along with anecdotal reports of bumblebees being seen in Victoria and New South Wales, suggests there may well have been illegal transfers of bumblebees to the mainland. However, the absence of widespread sightings indicates that perhaps there is some impediment to their survival there, maybe the same impediment that prevented them from establishing when first introduced over a hundred years ago. It appears that bumblebees will not be used in Australia, but the prospect of having them, then being thwarted, appears to have stimulated more research into managing native stingless and solitary bees on a commercial scale.[10]

Bombus hypnorum in Britain

Britain's situation is very different from our countries down in the Pacific. She has a rich fauna of native bumblebees, and nearby continental Europe is a constant potential source of new species, either by chance or design.

Around 2001 Britain was found to have acquired a new bumblebee species, *B. hypnorum* (the tree bumblebee), from continental Europe, where it is common.[11] As with the arrival of *B. terrestris* in Tasmania, the means of travel seems unknown and may have been deliberate or accidental. Unlike with Tasmania, wind-assisted free flight of queens was a possibility. But whereas the new bumblebee arriving in Tasmania (with no native bumblebees) was regarded as an unwanted alien, the attitude to *B. hypnorum* in Britain seems to have been more accepting. While it is considered an 'invasive species' in the scientific literature, the Bumblebee Conservation Trust website states that 'to have them nesting on your property is a real treat!'[12]

European subspecies of Bombus terrestris in Britain

While the arrival in Britain of *B. hypnorum* may have been a natural event, there have been deliberate commercial importations of non-native bumblebees. The native British buff-tailed bumblebee is *B. terrestris*, subspecies *audax*. The imported colonies from European commercial suppliers have mostly been of the continental European subspecies *B. terrestris dalmiticus* or *B. terrestris terrestris* (or hybrids between them).[13] Queens and males of the commercial subspecies have inevitably dispersed into the countryside since the initial trade in the 1990s and appear to have established as a resident species. Eventually, this became a conservation issue. Would foreign *dalmiticus* outcompete native *audax*?[14] Or would they interbreed and dilute the genetic integrity of *audax*? In 2015, based on the perceived risks, Natural England prohibited the use of non-native subspecies unless a special licence was obtained (for research or out of economic necessity).[15]

The Natural England proposal for the restriction acknowledged that there might be some cost increase to growers using the native *audax* but did not seem to support claims of poorer pollination performance. However, the British Tomato Growers' Association embarked on a study of *audax* performance and concluded that colonies of that subspecies often declined in strength soon after being installed in the greenhouses.[16] The study did not appear to have made side-by-side comparisons of the pollination strength of the different subspecies. Still, I find the alleged poorer performance of *audax* to be credible because the subspecies were rejected as a mass-rearing candidate early in the industry.[17]

We in New Zealand have no choice but to use *B. t. audax*. Even though after over a century here, it will not be identical to the current British population, the colony development results from our early exports to Koppert suggest New Zealand *audax* also have shorter colony durations than some continental European subspecies. Despite this apparent handicap, we have an economically viable greenhouse pollination industry – it is just that expectations and customer support are in line with the fewer weeks of colony activity. This tells me that other countries can have economically viable greenhouse pollination services without depending on the high-strength, long-lived bumblebee colonies based on particular European subspecies.

Another consequence of the apparently weaker colonies of *B. t. audax* is that if the New Zealand population has smaller colonies than the continental European species, the New Zealand commercial pollination suppliers need to be cautious in following their European support companies' hive stocking rate recommendations, especially for outdoor crops.

The extinction of *Bombus subterraneus* in Britain and its status in New Zealand

Commonly called the short-haired bumblebee, *B. subterraneus* was relatively common in Britain at the beginning of the twentieth century and figures in Sladen's classic 1912 book, but had disappeared from there by the end of the century. It was among the species introduced to New Zealand in 1905 but has remained restricted to a small region of the South Island. There were attempts early this century to reintroduce it to Britain from the New Zealand stock. That rearing exercise was unsuccessful, and so stock was obtained from Sweden. Colonies were produced, but it does not appear the species has established. Dave Goulson's book describes the project in detail.[18]

I visited Lake Tekapo where *B. subterraneus* is quite common, to take photographs for this book, during the early summer of 2021. I believe there are hints at the reasons for its very localised occurrence here in New Zealand and perhaps for its extinction in Britain. It may be an informative case study.

Figure 8.1 *Bombus subterraneus* on viper's bugloss near Lake Tekapo. **Photo credit Mies de Monchy.**

B. subterraneus is a moderately long-tongued, late-emerging, underground-nesting species. Sladen mentioned that queens produced a large first brood of workers, but colonies changed to queen production early and seldom reached a large size. He wrote, 'it does well in a short season'.[19] The large first brood is also a feature of an arctic species, *B. polaris*, where it seems to be an adaptation for a short colony cycle.[20]

In New Zealand, it is only common in a dry, high-altitude area with few flowers until very late in the spring, when viper's bugloss *Echium vulgare* bursts into bloom in vast quantities. I believe its lateness of emergence from hibernation suits it for this environment. Its short colony cycle is probably also an advantage, allowing it to produce new queens before the end of the short flowering season. There is no reason to think it 'likes' arid plateaux – when it was still common in Britain, it was apparently a lowland species. Its disappearance from Britain may be due to changing landscapes which have overtaken the habitat of a short but abundant flowering season in an otherwise near-barren landscape. During attempts to reintroduce the species to Britain, I did communicate my opinion to Dave Goulson that interspecific competition with earlier-emerging species may be a limiting factor. Hence, a dearth of early flowers may be an advantage. Dave did not disagree but pointed out that gaining public support for *removing* flowers would be difficult.

The *B. subterraneus* pattern is repeated on a less extreme scale by the distribution of *B. ruderatus* in New Zealand. This long-tongued species is also late emerging (but not as late as *B. subterraneus*) and, as observed by Lou Gurr, seems to be most prevalent where there is a lack of early spring forage, thus disadvantaging the early-emerging *B. terrestris*.[21] Others have also drawn attention to the tendency for rare bumblebee species to be late-emerging and potential victims of competition from earlier-emerging species.[22]

One conclusion from these observations is that interspecific competition may be a significant factor in bumblebee abundance and therefore that environmental enhancements of a general nature, such as a proliferation of flowers throughout the season, may only benefit species that are already common and may even be counter-productive for rare species, especially late-emerging ones such as is the case for several rare species in Europe. I also suggest that climatic correlations with a species' distribution should not necessarily be interpreted in terms of its physiological requirements (e.g. *B. subterraneus* does not seem to be adapted to low temperatures) but in terms of the local phenology and competing species.

The population crash of *Bombus occidentalis* and related species in North America

Whereas the decline of *B. subterraneus* in Britain appears to have been gradual and perhaps explicable in terms of its narrow ecological niche – it becoming outcompeted by other species – several species in North America seemed to have plunged in numbers from very common to rare within a few years.[23]

I need to explain bumblebee classification here. They all belong to the genus *Bombus*, within which there are several *subgenera*. The table below shows examples of two subgenera, subgenus *Bombus* (sometimes written 'Bombus s str' for *sensu stricto*), and *Pyrobombus*. The declining species are underlined. E NA, and W NA mean East and West North America, respectively.

Genus	Subgenus	species	subspecies	Location
Bombus	Bombus s str	*affinis*		E NA
		terricola		E NA
		occidentalis		W NA
		franklini		W NA
		lucorum		UK/Eur
		terrestris	*terrestris*	Eur
			dalmiticus	Eur
			audax	UK
	Pyrobombus	*impatiens*		E NA
		vosnosinskii		W NA
		hypnorum		UK/Eur
		jonellus		UK/Eur

Several of the subgenus *Bombus s str* have been amenable to domestication and produce large colonies. They have relatively short tongues and are the only group that are also 'nectar robbers,' and readily bite holes in flowers which are otherwise too deep for them to access the nectar.

Within the USA and Canada, several of this subgenus have declined in numbers from early this century. *B. franklini* (Franklin's bumblebee) is feared extinct, although it was not abundant before. But *B. occidentalis* (western bumblebee) in the West and *B. affinis* (rusty patched bumblebee) and *B. terricola* (yellow banded bumblebee) in the East were all common species. Now they are hard to find in many places. I have an interest in this group. *B. affinis* was a familiar, large and attractive species when I was a graduate student in Canada, and we reared it in Chris Plowright's lab. Rick Fisher, a student peer in Canada, and later research colleague at Massey University, reared *B. terricola* and studied its social parasite, *B. ashtoni*, for his PhD.

Figure 8.2 *B. affinis* (left) and *B. terricola* (right). (Specimens from my time in Toronto, later 1970s, when both were common).

Why did these formerly common bumblebees disappear? Global warming, pesticides and pathogens have been suggested as contributing factors.[24] There is also evidence that some pathogens and pesticides, while not serious individually, may have a harmful synergy in combination.[25] The puzzle is why this subgenus seems to have been differentially harmed while other bumblebee groups remain common. I believe this points more towards a pathogen than a broader-spectrum effect like pesticides (alone) or climate change.

More to point: what can be done to reverse the decline? Various jurisdictions have classified these species as 'endangered', which triggers various protections, mainly around pesticide use and habitat modification.[26] But where other bumblebee species remain common, I see no reason that these measures will significantly help the endangered species if they are declining for other reasons, such as a pathogen.

Captive rearing is on the list of remedial measures, which is important, not least to gain material to investigate whatever pathologies may be operating.[27] A comprehensive 2020 report describes the plans and procedures for 'ex-situ' work with *B. affinis*, which includes captive-rearing methods.[28] Wildlife Preservation Canada, a private organisation, has been rearing *B. terricola* from wild-caught queens, and I have an involvement with this project. The even more endangered *B. affinis* was too rare for any queens to be found.[29] This work should have the double advantage of enabling research on putative pathogens and other relevant aspects of the species' biology and providing an insurance population, thus providing material for re-release.

Reflections on bumblebee conservation

I selected the examples above based on personal interest and to show the variety of situations in which people may wish to preserve or eliminate bumblebees. The Australian example shows how the preservation of native biodiversity overrode the interests of the horticulture industry. These priorities were surely reversed a hundred years earlier when New Zealand was populated with British bumblebees for red clover pollination. The biodiversity-economic conflict is currently playing out in Britain with respect to what some may consider to be a fine point over the genetic composition of the local population of *B. terrestris*, while a new species, *B. hypnorum*, arrived and established with little fanfare. The sudden decline of a group of closely related but geographically dispersed species in North America has triggered a variety of conservation measures, which may or may not rescue those species.

Ecosystem services and biodiversity

The concept of 'ecosystem services' (ES) ranges from wild native plants being pollinated by wild native pollinators in an unmodified habitat, to commercial crops being pollinated by commercially managed honey- or bumblebees. There is strong interest in modifying habitats to increase the value of the pollination ES provided by wild bees, including bumblebees. The extent to which wild bumblebee populations (whether or not enhanced by habitat modification) contribute economically to crop yields will vary according to the 'numbers game' outlined in Chapter 3. Research in Wisconsin apple orchards showed that wild bees (bumblebees and solitary bees) sometimes provide the bulk of the pollination and may have a greater benefit than introduced honeybees.

And in Pennsylvania, the wild population of B. impatiens (common eastern bumblebee) was found to be effectively pollinating commercial pumpkin crops.[30] Another study reported a benefit over years of increased species diversity of wild bees for pollinating watermelons and highbush blueberries.[31] But these examples may not be typical of commercial crops. When I first collaborated with David Pattemore on the management of bumblebees for kiwifruit pollination here in New Zealand (Chapter 7), he was thinking and working in terms of the ES supplied by the local population of bumblebees, which his team was trying to enhance by providing additional nesting sites. We soon shifted the emphasis to managed rearing, concluding that it was impractical to boost the wild population to the extent that would be economically useful to kiwifruit growers. The 'numbers game' indicated that more intensive pollinator management would be required for that crop.

Some crops in some locations appear to yield well with the ES of the local wild pollinator populations, others benefit from local enhancements of the wild pollinators, and some need intensive pollinator management: the introduction of bees (bumble-, honey- or other) from outside. I found a paper titled 'Delivery of crop pollination services is an insufficient argument for wild pollinator conservation' to be persuasive.[32] It is not an anti-conservation paper. But it argues that the strategies to improve pollinator availability for crop production differ from those to promote threatened species. They point out that although wild bees contribute significantly to pollination services, the responsible species tend to be common and whose numbers can be increased by relatively low-cost interventions.

Implications for farming

The importance of biodiversity for agriculture is hard to dispute when framed in general terms, or when positive effects are listed.[33] But there are clearly individual situations where biodiversity must be subordinate to productivity.

Nearly half of Zonda Bees' income was from sales of the whitefly parasitoid Encarsia. These were a 'green' solution to a greenhouse pest, and their production exemplified how most farming is both 'ecological' (resource supplies, reproduction and survival rates drive the population size) and 'unnatural' (resources and mortality factors are manipulated). Encarsia prolificity was amazing – by providing the optimal environment for the host whitefly and shielding them from predators and parasites, we could grow vast numbers, and then with an appropriately timed 'infection' with the Encarsia – we were able to 'farm' vast numbers of natural enemies for whitefly control in greenhouses. Most farming is like this, whether growing crops, livestock or fish – optimise the physical environment and nutrition, eliminate enemies, and thus exploit the natural growth and reproductive potential of the product, most progeny of which would usually perish in nature.

While some interplanting of complementary crop species can be done on a small scale, most crops are grown as monocultures.[34] Biodiversity is enhanced on the crop margins with wildflower strips, hedgerows, riparian planting, etc. If such landscape modifications increase the local wild pollinator populations to a useful level, I am all for them. Despite the criticism of 'industrial' farming, it *is* an industry, and costs incurred in biodiversity enhancement must be borne by someone. I must confess that coming from a country with one of the lowest levels of taxpayer subsidy for agriculture (compared with the UK, EU, US, and Canada), I am very conscious that *someone has to pay* for eco enhancement.[35]

Future priorities

I want to encourage three areas of work: preservation of those bumblebee species deemed in danger of extinction, research directed at pollinator biology and management, and an entrepreneurial approach to new technologies.

Preservation of endangered species

I believe it is a misleading marketing ploy to equate the preservation of rare bumblebee species with the security of food production, but there are less utilitarian reasons to promote them. From an environmental ethics point of view, species have a 'right to exist'. And technologies and biological insights from bumblebee preservation projects may subsequently have wider practical applications.

I do not believe we fully understand the reason for the plummeting numbers of the *B. occidentalis – affinis* group of bumblebees in North America, but I suspect it is likely to be related to a pathogen. The decline of the iconic South American species *B. dahlbomii* is perhaps better understood – it seems most likely to be due to competition or pathogen transfer from the introduced *B. terrestris* and/or *B. ruderatus*.[36] If a wild population is endangered by disease or interspecific competition, there is no straightforward remedy. (There have been attempts to cull invasive *B. terrestris* in Japan, but it is difficult to see how this could be done selectively and at scale.)[37] I believe we must act to maintain protected populations in captivity. New Zealand has captive breeding programmes for several endangered birds and reptiles, and recently an insect. This is a significant function of zoos. Due to introduced mammalian predators being a severe threat to native wildlife here, we have several predator-proof reserves (up to several square kilometres in area) where native species flourish. More modest-sized screen cages could be valuable spaces for endangered bumblebees. Colonies could be initiated in rearing apparatus and liberated in protected enclosures, which would also serve as arenas in which to research other aspects of their biology.

Research

While there are numerous studies on bumblebees and flowers, their nesting sites are also important but are more challenging to investigate. I suggest that a modified version of the flower-nectar supplementing trick (page 16), whereby a full-up worker flies directly back to its nest, could be exploited in suitable habitats to locate wild colonies. This could be modified by making artificial feeding stations (with natural flowers used for training) to which wild foragers would commute. The domicile designs described in Chapter 1 could be worth building in large numbers to investigate nest site occupation. I have contemplated making a concrete false-underground domicile with walls thick enough to incorporate a tunnel.

The small size of bumblebee colonies – a maximum of a few hundred workers, compared to tens of thousands in a honeybee colony – is the Achilles' heel of managing them for pollination. The strength of commercial colonies is usually expressed as the number of adult workers present at point of sale. It is often assumed that around a third of the workers may be foraging at one time, but worker numbers can change rapidly (Figure 3.18) and it would be useful for forager trapping vestibules (Figure 3.7) to be used more widely.[38] This might settle some of the disparities between the apparent strength of different subspecies, such as *B. terrestris audax* (in New Zealand and Britain) and the commercial subspecies, and between conflicting recommendations for stocking rates (pages 68 and 71).

The commercial rearing companies have chosen (and genetically selected) a few species that produce large colonies, but it should be possible to manage other species in ways that maximise their colony sizes. I believe there are subtle factors which I would call 'socioeconomic' – the interplay of food supply, workload and queen-worker dominance relationships. My student Colin Tod obtained some tantalising results that could point to colony-maximising interventions.[39]

Bumblebee production consumes tens of tonnes of fresh-frozen, honeybee-collected pollen annually. It needs to be sterilised to prevent the transfer of honeybee diseases, and it is a major cost in bumblebee production. Pollen supply is vulnerable to problems in the honeybee industry, and bumblebee production could benefit from an affordable pollen substitute. There are such products for use by honeybee keepers, but honeybees feed their larvae at least partly on glandular secretions – the raw food material is reprocessed by the nurse bees, during which the biochemistry is undoubtedly modified. Bumblebee larvae are fed nearly pure pollen, so a more equivalent substance would be required. It does not need to be an exact synthetic copy of pollen – it needs optimal nutrients, appropriate fibre or particle content, and the phagostimulants that attract the bees to use it. There are great bioassays to test bumblebee food: small worker groups readily lay eggs and raise brood, and it is easy to set up numerous test replicates.[40]

An essential requirement for much of this research will be a supply of bumblebee colonies. So far, many projects have used commercially-produced colonies, but this limits the species range. Several research groups successfully rear bumblebees, but I believe most still use pollen lumps as an egg-laying substrate, which adds to the labour and skill demands compared to the starter-cup system described in Chapter 5.[41] Small-scale commercial production of this or the Bumblebox™ system could enable more efficient colony production for research.[42]

Technical innovation

The technology of pollination seems to seems to receive less attention than the ecology. Indeed there are highly technical developments around mechanical pollination, but I am thinking of the management of pollinating insects.[43] Prior to bumblebees being domesticated on an industrial scale, I expect many experts would have been sceptical that it was possible or at least economical. Similarly, when we contemplate almost any modern technological device, it is difficult for most of us to picture how it was designed and created. As I type this, I take it for granted that I can swap between Times New Roman and Arial fonts with ease, but selectable fonts were a radical innovation when introduced on the Apple Macintosh computer, inspired during Steve Jobs sitting in on a calligraphy class after dropping out of Reed College.[44] My point is that many technologies initially seem to defy common sense or what is 'possible,' and they need a rebellious spirit to push them. (They also tend to need financial support and the talents of a multidisciplinary team.)

A more high-tech approach could be applied to commercial-scale rearing of non-Bombus bees: stingless bees and various types of solitary bees. Some of this is well underway, and some species were domesticated years ago.[45] Rearing novel bee species requires a detailed understanding of their natural history and may involve disrupting or deleting some steps in the life cycle. The first approach should ask, 'What does the bee need?' The answer might point the researcher to an illuminated warehouse with robotic artificial flowers. This is just a wild 'arm waving' suggestion, but I feel it is a mistake to think too austerely.

These putative developments are literally 'biotechnology', but that term appears to be usually seen as the realm of microbiology, genetics and biochemistry. The linking of engineering with the life history of insects seems to fall outside current academic silos. These projects require a multi-disciplinary approach with commitments from biologists, chemists and engineers. Considering that a honeybee pandemic could devastate food production, perhaps some of the billions of pounds/euros/dollars spent to support primary industries could be directed at a large coordinated programme for managing alternative pollinators.

Use of heating cables

Use of heating cables

Although I have not written this book in the form of instructions, the heating cables used in the observation hives and starter cups may be challenging to use without additional information.

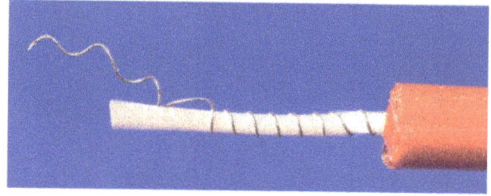

Figure A1. Silicone-insulated heating cable.

Figure A1 shows a partly stripped heating cable. The metal element is coiled on a glass-fibre core and coated with heat-resistant silicone rubber. They can be joined to normal copper cable with all-metal crimps or by wrapping the pulled-out wire around the copper cable, and coating with solder. The joint should be insulated with heat-shrink.

The electrical resistance of such a cable is proportional to its length, and in proprietary applications (e.g. heat pads, electric blankets, pipe de-icing) the length is fixed to give appropriate heat output for the voltage to which it is connected. The formulae below allow calculation of an appropriate length and resistance rating in DIY applications. The heating cable *must* be used in conjunction with a thermostat. The heat-power in watts *will not* equate to any fixed temperature.

DIY applications of these cables should use a low voltage. Twelve volts is easily obtained from computer power-supplies. (12v AC from an old-style transformer works equally well). Although this voltage is safe with respect to electric shock, it is quite possible to create excessive heat and fires if incorrectly applied (note the heat created if a 12 v car battery is shorted). All heating cable circuits should include a fuse (rated around twice the calculated normal current draw), and a thermal cut-out device (e.g. thermal fuse rated at 50 - 70°C). To avoid the cable overheating and melting or burning the insulating sheath, you need it to stay within a limited value of watts per metre. I suggest not exceeding 15 watts per metre (4.5 W/ft) for silicone insulated cables. Some cables are insulated with PVC plastic which is less heat resistant – PVC cables should not exceed 3w/m. Cable suppliers should provide maximum safe ratings for their products.

The total wattage required depends on the size of the object, its thermal insulation, and how cold the surroundings are. I suggest 3 – 8 watts for an observation hive of the type shown in Figure 2.8, and 10 – 12 watts per metre (3 – 4 W/ft) for rows of starter cups of the type shown in Figure 4.8.

Examples

P = power (W), V = volts, L = length (m), R = resistance (Ohms), m = metres, R/m = Ohms per metre. The calculations work the same if feet are substituted for metres in all expressions. I use 12 volts in the examples.

1.   Observation hive: purchasing cable of selected R/m:

1.1   Find the minimum length for the cable:

5 W heating for an observation hive; cable maximum heat (P_{max} / m) is 15 W/m
L_{min} = P_{tot} / (P_{max} / m) = 5 / 15 = 0.33 m.

Therefore the cable must be longer than 0.33m (which is inconveniently short to wrap around a hive base anyway). It does not matter how *much* longer the cable is – if selected by the criteria below, it will still provide 5 W of heating power.

1.2.   Choose a cable length, and find the required resistance rating. A longer cable spread over the hive base will proved more dispersed warming. Say, choose 1.5 m.

R/m = (V^2 / P) / L = (12^2 / 5) / 1.5 = 19.2 Ohm/m. Choose 20 Ohm/m cable.

2.   Observation hive: you have access to PVC insulated, 60 Ohm/m, cable from an electric blanket. How can it be used?

2.1   Find the minimum length for the cable (max W/m is 3 for PVC insulation):

Lmin = P_{tot} / (P_{max} / m) = 5 / 3 = 1.66 m.

Find the heat output from this length:

P = V^2 / (L x (R/m)) = 12^2 / (1.66 x 60) = 1.66 W This is much less than 5.

Solution: Use two 1.66m lengths in parallel. They will each provide 1.66 W. x 2 = 3.33 W which is within the range 3 – 8 and will be enough to heat a well-insulated hive. Using the same logic some cables may be satisfactory with three or more lengths in parallel. Note: using a longer cable will produce less heat.

3. Rows of starter cups.

The length of the cable is fixed by the width of the row of cups. This system has greater heat loss than an insulated observation hive, and requires 10 – 12 watts per lineal metre of cups (3 – 4 W/ft), so there are fewer cable choices. As mentioned in Chapter 4, a fully insulated and earthed mains voltage soil warming cable of 60 W is suitable for around 6 m (20 ft) of cups. Be aware that some warming cables have built in thermostats that limit the temperature well below what is required for bumblebees, and need to be by-passed.

To find the resistance rating use the formula:

$R/m = (V^2 / P) / L$

The ratings I use are 2 Ohm/m for a double row of cups, 2 x 1.2 m (actual cable length 2.6 m to pass between the rows). For a single row I use 5 Ohm/m. All on 12 v.

All arrangements of the formulae:

$P = V^2 / (L \times (R/m))$

$R/m = (V^2 / P) / L$

$L = (V^2 / P) / (R/m)$

$P/m = V^2 / ((R/m) \times L^2)$

$P/m = P / L$

$L_{min} = P_{tot} / (P_{max} / m)$

Contact me - nelsonpomeroy@gmail.com - for direct assistance.

Bumblebee genetics and diploid males

Bumblebee genetics and diploid males

Genetics 101

Our bodies are made up of microscopic cells, and each cell contains an identical set of chromosomes, which come in pairs (*diploid*). Humans have 26 pairs, AA, BB, CC,…XX, in women. Men have AA to WW, and an odd 'pair', XY. Sex cells (ova and sperm) result from a separation of the chromosome pairs into singles (*haploid*). So each ovum contains one each of the chromosomes A – X. Sperm cells contain one each of chromosomes A – W, plus *either* one X or one Y. After fertilisation you get pairs again. If an X sperm combines with an ovum the result is XX, a female. If a Y sperm combines with an ovum the result is XY, a male. This pattern, or an equivalent of it, is found in most creatures including most insects.

Chromosomes are long strands of DNA. Sections of DNA which code for specific traits (like eye colour) are called *genes*. The gene for eye colour can have different forms called *alleles*, for example blue or brown. (Fact alert: human eye colour is more complicated than this but it fairly illustrates the principle.) A person may have a brown allele on one chromosome of a pair, and a blue allele on the other chromosome (*heterozygous*) or the two alleles may be both brown or both blue (*homozygous*).

Sex determination in bees, wasps and ants

The evolutionary predecessors of bees, wasps and ants were thought to be parasitoids that laid their eggs inside a living host (victim) creature. Parasitoid females tend to be bigger than males, and since the host might be of variable size, it should be beneficial to lay a female egg in a big host and a male egg in a smaller host. But this required control over egg sex, which is not possible in the above pattern of random assortment. So, it is hypothesised, they gained the ability to chose whether to fertilise an egg or not, and that fertile ones would be female and unfertilised ones would be male (most insects store sperm after mating and it is released to the ova at the point of egg laying).[1] This has carried through to modern bees.

The ova of bees, like of most creatures, are the result of meiosis and contain a single set of chromosomes. So if they are not fertilised by a sperm cell the resultant bee is stuck with a single set of chromosomes (remains haploid) instead of the usual two, which female bees have. On the face of it, it appears that single chromosomes equal male; double chromosomes equal female. Remember that workers are also female.[2]

Diploid males

But sometimes male bumblebees have been found that have two sets of chromosomes: *diploid males*. It is thought to happen like this: femaleness in bees requires a specific 'sex' gene to be heterozygous – to have non-matching alleles, such as the example above of a person having non-matching brown and blue alleles for eye colour. The sex gene in bumblebees contains several tens of different alleles, and the allele of a random male in the field is not likely to be the same as a particular queen who mates with him. In Figure A2 I have used different colours to represent different alleles on the sex chromosome.

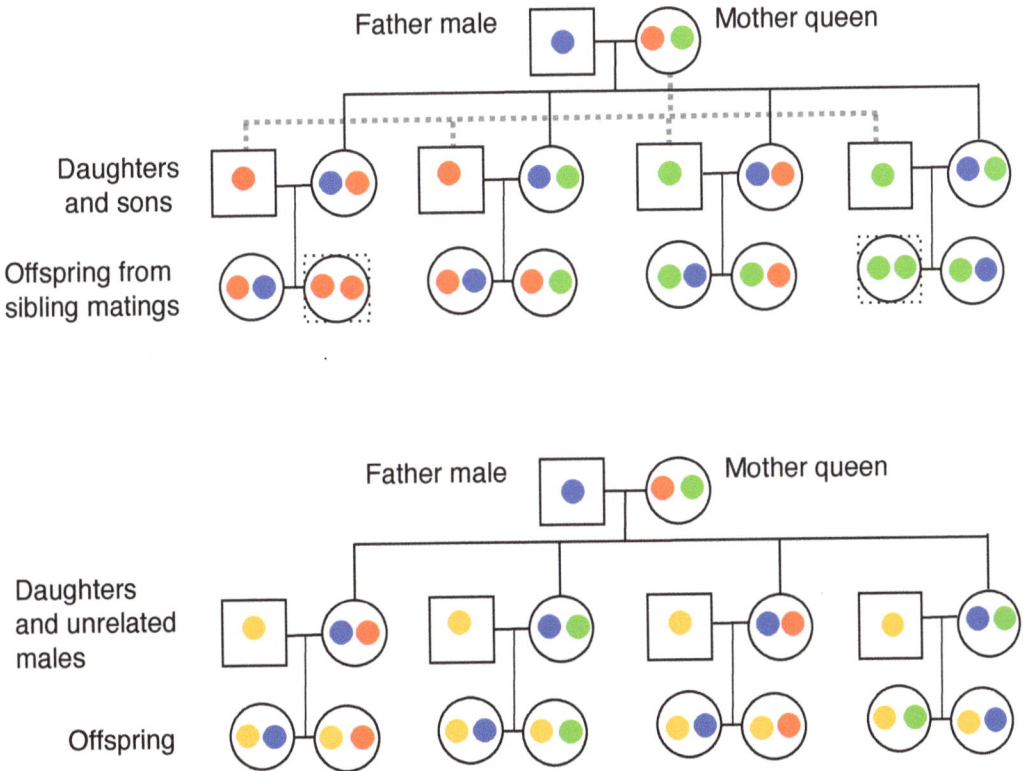

Figure A2. Pedigree diagrams.

Upper: (The thick dotted line leading to the males is because they are not the result of a cross between the parents – they arise from unfertilised eggs from the queen only.) Young queens mate with their brothers, and in two of the four combinations half of the fertilised offspring are homozygous for the sex allele (dotted squares to indicate these will be physically male).

Lower: The same young queens mating with unrelated males: all of the female offspring are heterozygous. If there are several tens of 'colour' options, the chance of them matching in the fertilised offspring is very small.

Acknowledgements

Acknowledgements

My older brother Grenville, home from teachers' training college, told me how to make an insect collection when I was eight years old (such things were still encouraged in 1960). We never know 'what would have been' but that was followed by a lifelong enthusiasm with insects. My parents, Joyce and Howard Pomeroy, were benignly tolerant of the domestic disruptions from the various wildlife, alive and dead, that I brought home. My father taught me macro photography.

At Massey University, I was imbued with a sense of the animal's point of view by animal behaviour specialist Lou Gurr. My erratic undergraduate record almost barred me from graduate studies but John Skipworth pushed for my admission, apparently declaring I was 'an enthusiast'. My fortuitous admission to the University of Toronto, detailed in Chapter 2, was enabled by Chris Plowright, who showed me the virtue of controlled experiments. I think my experience in the more financially-liberal North American environment primed me for a more ambitious approach to later research.

My subsequent commercially-oriented work at Massey University was fully encouraged by Brain Springett. Wrightson Horticulture was the major funder and enabled us to expand floor space and other resources beyond the typical capacity of the Department of Botany and Zoology. Additional funding came from the New Zealand Kiwifruit Authority and the E and C Thoms Trust. The financial support enabled the employment of a full-time technician, Stan Stoklosinski, who was an active contributor to the design and construction of rearing equipment, and data collection. Rick Fisher collaborated on various pollination projects. Over the years the other part-time and short-term assistants are too numerous to name, but you know who you are.

The New Zealand visit of Henri Oosthoek and Adriaan van Doorn from Koppert BV in the Netherlands was another big turning point and I had the privilege of being involved in the 'ground floor' of the new industry of bumblebee production (as detailed in Chapter 5). Koppert's funding was significant for the University and for me personally. The subsequent decade of business partnership with Koppert as Zonda Bees showed me some realities of the business world.

Linda Cassells provided valuable advice during the early stages of writing. I am grateful to Jeremy Burbidge of Northern Bee Books for being prepared to take on this project and to Simon Paterson and Felicity Potter for their designing and editing inputs.

Single-minded enthusiasm for any activity can be a cost for the person's family. My respective wives Rachel and Mies lived through difficult periods because of my 'work', yet supported and participated in it in various ways. My children likewise.

Notes

Notes

Preface

1   Phiri, B.J., Fèvre, D. & Hidano, A. Uptrend in global managed honey bee colonies and production based on a six-decade viewpoint, 1961–2017. *Scientific reports* 12: 21298. DOI: 10.1038/s41598-022-25290-3

Chapter 1. Bumblebees in the Field

1   The farm straddled the Moumahaki Stream, Barrow Road, and was bordered by Omahina Road and the Omahina Stream. Our house was at 39°43'55.8" S, 174°41'05.1" E. It was demolished around 2000 to recycle the native timbers. The current Google Earth (2022) image shows the shadow of the brick chimney.

2   Hopkins, I. 1914. History of the bumblebee in New Zealand: Its introduction and results. *New Zealand Department of Agriculture, Industry and Commerce* 46: 1-29.

3   Rasmont, P., Coppee, A., Michez, D.and De Meulemeester, T. 2008. An overview of the *Bombus terrestris* (L. 1758) subspecies (Hymenoptera: Apidae), *Annales de la Société Entomologique de France* 44: 243-250. DOI: 10.1080/00379271.2008.10697559

4   Cumber, R.A. 1949. An overwintering nest of the humble-bee: *Bombus terrestris* (L.) (Hymenoptera, Apidae). *New Zealand Science Review*. 7: 96-7.

Gurr, L. 1973. Evidence of overwintering nests of *Bombus hortorum* and *Bombus ruderatus* (Hymenoptera: Apidae) in New Zealand. *New Zealand Entomologist* 5: 339-41.

Cumber excavated a colony which was producing new queens in November (Spring), and assuming they would not normally be produced until the end of Summer, concluded the colony had 'overwintered' i.e. survived from the previous Summer. Gurr formed the same conclusion on the basis of finding males in the field in November. My observations indicate that colonies of *B. terrestris* established in late Winter or early Spring frequently produce new queens in November, which I see as a normal colony cycle. Bumblebees do not seem to be tied to the seasons in a direct way. Queens establish colonies when the weather permits flight, and there are enough flowers. Then the colonies grow and senesce over a period often of a few months.

5   Sladen, F.W.L. 1912. *The humble-bee. its life history and how to domesticate it.* London: Macmillan, 283 pp.

Free, J.B. and Butler, C.G. 1959. *Bumblebees.* Collins, London.

6   Pomeroy, N. 1981. Use of natural sites and field hives by a long-tongued bumble bee *Bombus ruderatus. New Zealand Journal of Agricultural Research* 24: 409-414. DOI: 10.1080/00288233.1981.10423408

7   Some foragers sometimes stay out overnight, but night time is clearly the best time to get most of them.

8   Bumblebees have a physiological block to flying in darkness. If a bumblebee is flying under electric lighting and you flick the light off, the bee drops to the floor. It is well known that the visual spectrum of insects is shifted towards shorter wavelengths than human vision (they see ultra-violet and are less sensitive to red) and the use of red light when working with bumblebees to inhibit them flying is a long established practice. However, I have recently found *B. terrestris* workers will fly under very bright (1W) pure red LEDs (peak intensity around 620 nm).

9   Free and Butler 1959, see note 5.

10   Donovan, B.J. and Wier, S.S. 1978. Development of hives for field population increase, and studies on the life cycles of the four species of introduced bumble bees in New Zealand. *New Zealand Journal of Agricultural Research* 21: 733-756

11   Gurr L. 1972. The introduction of bumblebees into North Island, New Zealand, *New Zealand Journal of Agricultural Research* 15: 635-638. DOI: 10.1080/00288233.1972.10430553

12   The upholsterers' cotton locally available at the time was a blue-grey mixture of fibres recycled from old textiles, and was used as padding in furniture. I am not sure if it was all cotton although synthetic fibres were less common at the time. The mouse nests were made by a strain of nest-building mice from Massey University's Small Animal Production Unit. I initially gave them plain upholsterers' cotton which would have provided a more direct comparison with unscented cotton, but they seemed to have difficulty making nests with it: it became matted and trampled, hence mixing it with hay.

13   Sladen 1912, see note 5. Page 112

14   Lye, G.C., Park, K.J., Holland, J.M., and Goulson, D. 2011. Assessing the efficacy of artificial domiciles for bumblebees. *Journal for Nature Conservation* 19: 154-160. DOI: 10.1016/j.jnc.2010.11.001.

Johnson, S.A., Tompkins, M.M., Tompkins, H., and Colla, S.R. 2019. Artificial domicile use by bumble bees (*Bombus*; Hymenoptera: Apidae) in Ontario, Canada. *Journal of Insect Science* 19(1): 7. DOI: 10.1093/jisesa/iey139.

Lye *et al* 2011 do not provide enough information for me to visualise the entrance appearance. It is extremely difficult to terminate a plastic tube flush with the ground surface: if it ends below the surface it will tend to fill with earth, and if it end above it does not have the same topography as a burrow in the ground. They attempted to provide a more natural entrance appearance by covering the entrance to the tunnel with a tile, but it is still difficult to know whether the tube projected out of the soil. I note also that the diagrams show the tube projected into the nest cavity. Depending on the arrangement of nest material this could also be difficult for a queen to navigate.

The tunnels on the domiciles of Johnson *et al* 2019 were straight PVC pipes, resting on or suspended slightly above the ground.

Chapter 2. Bumblebees in the Labratory

1   Sakagami, S.F. 1966. Techniques for the observations of behavior and social organization of stingless bees by using a special hive. *Papéis Avulsos de Zoologia. Secret Agric. Sao Paulo* 19: 151-162.

2   Heinrich, B. 1979. *Bumblebee Economics*. Harvard University Press. 245 pp.
     Bumblebee metabolic output is twice that of hummingbirds, the most active vertebrates.

3   This design was published by: Pomeroy, N. and Plowright, R.C. 1980. Maintenance of bumble bee colonies in observation hives (Hymenoptera: Apidae). *The Canadian Entomologist* 112: 321-326. DOI: 10.4039/Ent112321-3

4   Sladen 1912, see Chapter 1, note 5. Page 97
     Free and Butler 1959, see Chapter 1, note 5.

5   Free, J. B. 1955. Queen production in colonies of bumblebees. *Proceedings of the Royal Entomological Society of London.* (A). 30: 19-25. DOI: 10.1111/j.1365-3032.1955.tb00164.x

6   Röseler, P-F. 1967. Untersuchungen uber das Auftreten der 3 Formen im Hummelstaat. *Zoologische Jahrbücher. Abteilung für allegemeine Zoologie und Physiologie der Tiere* 74: 178-197.

7   Hamilton, W.D. 1964. The genetical evolution of social behaviour. II *Journal of Theoretical Biology* 7: 17-52.

8   Plowright, R.C. and Jay, S.C. 1966. Rearing bumble bee colonies in captivity. *Journal of Apicultural Research* 5: 155-165. DOI: 10.1080/00218839.1966.11100149

9   Goulson, D. 2013. *A Sting in the Tail*. Jonathan Cape, London, UK. 256 pp.

Chapter 3. Bumblebees in the crop and orchard: the pollination numbers game

1   Donovan, B.J .and Wier, S.S. 1978. Development of hives for field population increase, and studies on the life cycles of the four species of introduced bumble bees in New Zealand. *New Zealand Journal of Agricultural Research* 21: 733-56.

2   Rod P. Macfarlane, personal communication.

3   Palmer-Jones, T., Forster, I.W. and Clinch, P.G. 1974. Observations on the pollination of Montgomery red clover (*Trifolium pratense* L.). *New Zealand Journal of Agricultural Research*, 9: 738-747. DOI: 10.1080/00288233.1966.10431563

4   Macfarlane, R.P., van den Ende, H.J., and Griffin, R.P. 1991. Pollination needs of 'Grasslands Pawera' red clover. *6th Pollination Symposium. Acta Horticulturae* 288: 399-404.

5   Pomeroy, N. and Fisher, R.M. 2002. Pollination of kiwifruit (*Actinidia deliciosa*) by bumble bees (*Bombus terrestris*): Effects of bee density and patterns of flower visitation. *New Zealand Entomologist* 25: 41-49 DOI: 10.1080/00779962.2002.9722093

    Pomeroy, N. 2001. The big bumble bee trial. *New Zealand Kiwifruit Journal*. 147: 35-39

6   Hopkins 1914, see Chapter 1, note 2.

7   Darwin, C.R. 1878. *The effects of cross and self-fertilisation in the vegetable kingdom*. John Murray, London, UK. 487 pp.

8   Gurr. L. 1964. The distribution of bumblebees in the South Island of New Zealand. *New Zealand Journal of Science* 7: 625-642.

9   Gurr, L. 1974. The role of bumblebees as pollinators of red clover and lucerne in New Zealand: A review and prospect. *Proceedings of the New Zealand Grasslands Association* 36: 111-120.

10 Hobbs, G.A. 1967. Obtaining and protecting red-clover pollinating species of *Bombus* (Hymenoptera: Apidae) in Alberta. *The Canadian Entomologist* 99: 943-951.

11 Macfarlane R.P., Griffin, R.P. and Read, P.E.C. 1983. Bumble bee management options to improve 'Grasslands Pawera' red clover seed yields. *Proceedings of the New Zealand Grasslands Association* 44: 47-53. This paper recommended additional early-flowering crops to help support the bumblebee population. Some red clover growers planted a broad bean crop and/or grew some early-flowering uncut red clover.

12 After forty years my records no longer show the area of the University's crop on which I placed 70 hives. Some notes suggest around 1.2 ha. I also do not have a record of the area of the crop on which no hives were placed, but I recall it was the largest, and I have estimated it was 1.5 ha. These uncertainties do not affect the stocking rate conclusions which were based on bee densities and seed production within small sub-plots.

13 Pomeroy, N. and Stoklosinski S.R. 1990. Measuring the foraging strength of bumble bee colonies. A "vestibule Trap". *11th International Congress – International Union for the Study of social Insects*. Bangalore, India. Page 469.

    Sixteen years later a paper was published with a 'new technique' for trapping bumblebees which used the same one-way principle but was not promoted as a measure of the standing crop of foragers, just a measure of foraging rate: Martin, A.P., Carreck N.L., Swain, J.L., Goulson, D., Knight, M.E., Hale, R.J.,

Sanderson, R.A. and Osborne, J.L. 2006. A modular system for trapping and mass-marking bumblebees: applications for studying food choice and foraging range. *Apidologie* 37: 341–350

14 Jing, S., and Boelt, B. 2021. Seed Production of red clover (*Trifolium pratense* L.) under Danish field conditions. *Agriculture* 11(12): 1289. DOI: 10.3390/agriculture11121289

15 Macfarlane *et al* 1983, note 11. This paper refers to *Bombus hortorum* colonies averaging '679 bees of which approximately 50% would have been actively working over the flowering period.' This (personal communication) was based on empty cocoon numbers after the colonies declined, with a formula for the allocation of small cocoons between males and workers. I was sceptical of these figures and suspected that the foraging population would be much smaller. Hence the desire for a more direct measure.

16 Macfarlane *et al* 1983, note 11.

17 Kilgour, M., Saunders, C., Scrimigeour, F. and Zellman E. 2007. *Kiwifruit. The key elements of success and failure in the NZ kiwifruit industry.* Lincoln University Research archives. https://researcharchive.lincoln.ac.nz/bitstream/handle/10182/3631/Kiwifruit.pdf;sequence=1

Wikipedia: https://en.wikipedia.org/wiki/Kiwifruit_industry_in_New_Zealand

Peacock, C. 2005. Kiwi Jackpot. *New Zealand Geographic.* Issue 071 https://www.nzgeo.com/stories/kiwi-jackpot/

18 Free, J.B. 1993. *Insect pollination of Crops.* Academic Press, London, UK. 684 pp. http://treefruit.wsu.edu/article/crop-load-management-back-to-basics/ Accessed 30/1/2023

19 Crane, J.C. 1969. The role of hormones in fruit set and development. *Hortscience,* 4(2): 8-11.

20 Broussard, M.A., Goodwin, M., McBrydie, H.M., Evans, L.J. and Pattemore, D.E. 2021. Pollination requirements of kiwifruit (*Actinidia chinensis* Planch.) differ between cultivars 'Hayward' and 'Zesy002', *New Zealand Journal of Crop and Horticultural Science,* 49(1): 30-40. DOI: 10.1080/01140671.2020.1861032

21 Ferguson, A. 1984. Kiwifruit: a botanical review. In: *Horticultural Reviews,* (Ed. J. Janick). DOI:10.1002/9781118060797.ch1

22 Ferguson 1984. See note 21.

23 Palmer-Jones, T. and Clinch, P.G. 1974. Observations on the pollination of Chinese gooseberries variety "Hayward". *New Zealand Journal of Experimental Agriculture* 2: 455-458.

24 Goodwin, M. 1987. *Ecology of the honey bee (Apis mellifera L.) pollination of kiwifruit (Actinidia deliciosa [A. Chev.]).* Unpublished PhD Thesis, University of Auckland.

25 Broussard *et al* 2021. See note 20.

26 A caveat to the use of trapping vestibule data: they may overestimate the number of foragers working the crop. I set the vestibules in three cages after visually counting the workers on the flowers, and trapped around twice as many pollen collectors in the vestibules as I counted on the flowers. That means that either I was failing to see all the foragers on the flowers, or the bees spent a significant part of their time outside the hive not actually working the flowers. Also, extrapolation of the numbers of foragers I counted in the open orchard to the whole 1.2 ha came to 329 in total, which is equivalent to far fewer mean foragers per hive than I was measuring in the vestibules. Again, perhaps my visual forager count were failing to see all the bees, or the vestibule counts may overstate the number on the flowers due to significant time spent outside the hive but not foraging.

27 Cutting, B.T., Evans L.J., Paugam, L.I., McBrydie, H.M., Jesson, L.K., Pomeroy, N., Janke, M., Jacob, M., and Pattemore, D.E. 2018. Managed bumble bees are viable as pollinators in netted kiwifruit orchards. *New Zealand Plant Protection* 71: 214-220. DOI: 10.30843/nzpp.2018.71.178

28 https://www.biobees.co.nz/pages/kiwifruit
https://www.zonda.net.nz/product/29/kiwifruit-actinidia-deliciosa/
Both accessed February 2023

29 https://www.biobestgroup.com/en/news/bumblebees%2C-the-better-alternative-for-kiwi-pollination Accessed February 2023

30 Céline Cazy, Zespri Group. Personal communication for access to the report.

Chapter 4. Towards bumblebee domestication

1   Velthuis, H.W. and van Doorn, A. 2006. A century of advances in bumblebee domestication and the economic and environmental aspects of its commercialization for pollination. *Apidologie* 37: 421–451

2   Myers, N., Mittermeier, R.A., Mittermeier, C.G. da Fonseca, G.A.B. and Kent, J. 2000. Biodiversity hotspots for conservation priorities. Nature 403: 853–858 DOI: 10.1038/35002501

Thomsen, T. 2021. *The lonely Islands: The evolutionary phenomenon that is New Zealand*. Reed New Holland, Auckland and Sydney.

3   https://treecrops.org.nz/tagasaste, accessed December 2022.

4   Heinrich, B. 1979. See note 1, Chapter 2.

5   Plowright, R.C. and Jay, S.C. 1966. Rearing bumble bee colonies in captivity. *Journal of Apicultural Research* 5(3): 155-165. DOI: 10.1080/00218839.1966.11100149

6   'Simmerstat™' (https://trademark.trademarkia.com/simmerstat-71503017.html)
    is a type of power control for hot plates and cookers. It contains a temperature-
    sensing bimetallic strip, wound with a tiny heating element. When power is
    applied, the strip heats up and bends, which switches it off – when it cools it
    switches on again. It is important to understand that this does not sense the
    temperature of the appliance itself, it is just a method of cycling the power on and
    off. The adjustment increases the proportion of 'on' time. My home made version
    included a heat-generating 5 W ceramic resistor glued to a heat-dispersing metal
    strip which was close to the heat-sensing strip of a fish-tank thermostat. This
    cycled on and off several times a minute depending on the thermostat setting.
    Quite separately I measured the temperature at the cloth knobs, and adjusted the
    cycle time by trial error until the knobs remained around 30°C (86°F). A system
    like this, that does not 'know' the target temperature, only remains accurate if the
    overall ambient temperature stays constant: It takes more energy to maintain 30°C
    knobs in a cool room than in a warm room.

7   Similarly to the principle in Note 6 above, but less extreme, it did not seem
    practical to attach the thermostat sensor directly inside a cup, due to the distance
    from the heat source, which would give an on/off lag (hysteresis). This is the
    same as with many electric frying pans where the sensing component is in a
    tube projecting from the plug – they have such a big response delay that they
    alternate between too hot and too cold. By controlling the metal strip to 32°C
    (90°F), while the room temperature was approximately 25°C (77°F) the knob
    temperature was close to the ideal 30°C (86°F).

8   When I kept fully-confined colonies in Chris Plowright's lab, anecdotal
    observations and a suggestion from an experiment in my thesis research there
    persuaded us that colonies would keep growing for longer if we removed
    a proportion of the workers, as would happen with the natural attrition of
    free-foraging colonies. The paper by Röseler (1967, see Chapter 2, note 4) also
    pointed in this direction.

9   Woodward, D.R. 1990. *Food demand for colony development, crop preference and
    food availablility for Bombus terrestris (L.) (Hymenoptera: Apidae)*. PhD thesis,
    Massey University. https://mro.massey.ac.nz/handle/10179/4588?show=full

    Tod, C. 1986. *Socio-economic effects on colony size in the bumble bee Bombus
    terrestris (Hymenopters: Apidae)*. MSc thesis, Massey University. https://www.
    researchgate.net/publication/332413194_socio-economic_effects_on_colony_
    size_in_the_bumble_bee_bombus_terrestris_hymenoptera_apidae

10 Meyer-Rochow, V.B. 2019. Eyes and vision of the bumblebee: a brief review on
    how bumblebees detect and perceive flowers. *Journal of Apiculture* 34(2):107-
    115. DOI: 10.17519/apiculture.2019.06.34.2.107

11   Massey University 1986. *Bumble bee colony initiation apparatus*. New Zealand Patent 216115, Complete specification. https://app.iponz.govt.nz/app/Extra/IP/Mutual/Browse.aspx?sid=638115292771038018 Accessed February 2023.

12   Milne, J.B. 2014. *The New Zealand kiwifruit industry - challenges and successes 1960 to 1999*. MA thesis, History, Massey University. https://mro-ns.massey.ac.nz/bitstream/handle/10179/5554/02_whole.pdf?sequence=2&isAllowed=y Accessed February 2023

Chapter 5. Bumblebees in greenhouses – Genesis of a unique industry

1   Fisher, R.M. and Pomeroy, N. 1989. Pollination of greenhouse muskmelons by bumble bees (Hymenoptera: Apidae). *Journal of Economic Entomology* 82(4):1061-1066

2   History of Biobest company: https://www.biobestgroup.com/en/pioneer

3   History of Koppert company: https://www.koppert.com/about-koppert/our-company/

4   Toni, H.C., Djossa, B.A., Tele Ayenan, M.A. and Teka, O. 2021. Tomato (*Solanum lycopersicum*) pollinators and their effect on fruit set and quality. *The Journal of Horticultural Science and Biotechnology* 96(1):1-13. DOI: 10.1080/14620316.2020.1773937

5   Velthuis and van Doorn 2006. See Chapter 4, note 4.

6   https://www.massey.ac.nz/study/scholarships-and-awards/j-p-skipworth-scholarship-ecology/ Accessed February 2023.

7   Velthuis and van Doorn 2006. See Chapter 4, note 4.

8   See Appendix 2 for an explanation of the effect of inbreeding and diploid males.

9   Röseler P.-F. 1985. A technique for year-round rearing of *Bombus terrestris* (Apidae, Bombini) colonies in captivity. *Apidologie* 16:165–170.

Chapter 6. Industry development and personal challenges

1   OECD farm support data: https://data.oecd.org/agrpolicy/agricultural-support.htm
       New Zealand Ministry for Primary Industries. 2017. New Zealand agriculture: a policy perspective. https://www.mpi.govt.nz/dmsdocument/27282-New-Zealand-Agriculture Accessed February 2023.

2   Acute oral LD50 for copper sulphate in rats is 500 parts per million, and twice that level (half as toxic) for most triazole fungicides. Sources: *Introduction and Toxicology of Fungicides*. National pesticides information center. http://npic.orst.edu/ingred/ptype/fungicide.html. Veterinary Medicines Directorate

UK Product information database. https://www.vmd.defra.gov.uk/ productinformationdatabase/files/SPC_Documents/SPC_107208.PDF Accessed November 2022.

3   Velthuis and van Doorn, 2006. See Chapter 4, note 4.

4   https://www.zonda.net.nz/

Chapter 7. After Zonda Bees: travel, school and kiwifruit again

1   Iwasaki J.M., Barratt B.I., Lord, J.M., Mercer, A.R., Dickinson, K.J. 2015. The New Zealand experience of *Varroa* invasion highlights. Research opportunities for Australia. *Ambio*. 44(7): 694-704. DOI: 10.1007/s13280-015-0679-z. Epub 2015 Jul 2. PMID: 26133152; PMCID: PMC4591231.

2   Pomeroy 2001. See Chapter 3, note 5.

3   Goodwin 1987. See Chapter 3, note 24.

4   The New Zealand Institute for Plant and Food Research Limited is a New Zealand government-owned Crown Research Institute. https://www. plantandfood.com/en-nz/about-us

5   Pomeroy, N. 2015. "Bumble bee production for pollination," Chapter in V. E. Neall (ed), *Plains' Science: inventions, innovations and discoveries from the Manawatu – 2*, Royal Society of New Zealand Manawatu Branch Inc. and The Science Centre Inc., Palmerston North, New Zealand. pp 112-121

6   Johnson, R.M., Ellis, M.D., Mullin C.A. and Frazier, M. 2010. Pesticides and honey bee toxicity – USA. *Apidologie*, 41(3): 312-331. DOI: 10.1051/ apido/2010018

7   Evans, L.J., Cutting, B.T., Jochym, M., Janke, M.A, Felman, C., Cross, S., Jacob, M., and Goodwin, M. 2019. Netted crop covers reduce honeybee foraging activity and colony strength in a mass flowering crop. *Ecology and Evolution*. 29: 5708-5719. DOI: 10.1002/ece3.5154. PMID: 31160992; PMCID: PMC6540661.

8   For information on the BumbleBox™ please email bumblebox@plantandfood. co.nz. The New Zealand Institute for Plant and Food Research Limited is a New Zealand government-owned Crown Research Institute.

Chapter 8. The environment and bumblebees, rejected or cherished

1   Goulson, D. 2012. *Bumblebees: behaviour, ecology, and conservation*, Second edition. Oxford University Press, Oxford, UK.

Hirsch, E.R. 2020. *Where have all the bees gone? Pollinators in crisis*. Twenty-first Century Books, Minneapolis, USA.

Packer, L. 2010. *Keeping the bees: Why all bees are at risk and what we can do to save them*. Harper Collins, Toronto, Canada.

2   'Social' insects are those which live together in cooperative groups of related individuals. New Zealand has social ants but only solitary native bees and wasps.

3   Sladen 1912, (See Chapter 1, note 5) p 159, stated that *B. terrestris* had become acclimatised on both Australia and New Zealand.

4   Schmid-Hempel, P., Schmid-Hempel, R., Brunner, P., Seeman, O.D. and Allen, G.R. 2007. Invasion success of the bumblebee, *Bombus terrestris*, despite a drastic genetic bottleneck. *Heredity*. 99: 414–422. DOI: 10.1038/sj.hdy.6801017

5   Buttermore, R.E., Pomeroy, N., Hobson, W., Semmens, T., and Hart, R. 1998. Assessment of the genetic base of Tasmanian bumble bees (*Bombus terrestris*) for development as pollination agents. *Journal of Apicultural Research*. 37: 23–25.

6   Schmid-Hempel *et al* 2007. See note 4.

7   Dafni, A., Kevan, P., Gross, C.L., and Goka, K., 2010. *Bombus terrestris*, pollinator, invasive and pest: An assessment of problems associated with its widespread introductions for commercial purposes. *Applied Entomology and Zoology*. 45(1): 101–113. http://odokon.org/

8   Hunter, M.N. 2007. *The future development of the Australian honeybee industry*. Submission to the House of Representatives (Australian Government), Agriculture, Fisheries, and Forestry Committee Inquiry. Centre for International Economics, Canberra & Sydney. https://www.aph.gov.au/parliamentary_business/committees/house_of_representatives_committees?url=/pir/honeybee/subs/sub056.pdf Accessed February 2023

9   Australian government 2017. *Risks and opportunities associated with the use of the bumblebee population in Tasmania for commercial pollination purposes*. Senate Environment and Communications References Committee. Canberra. https://apo.org.au/sites/default/files/resource-files/2017-06/apo-nid94421.pdf Accessed February 2023.

10   Hort Innovation 2022. *Development of blue-banded bees as managed buzz pollinators*. https://www.horticulture.com.au/growers/help-your-business-grow/research-reports-publications-fact-sheets-and-more/ph19001/ (Accessed January 2023).
     Halcroft, M., Spooner-Hart, R., and Dollin, A. 2013. "Australian stingless bees," in Vit, P., Pedro, S.R.M, and Roubik, D.W. (eds.), Pot-Honey *A Legacy of Stingless Bees*. Springer, New York, Heidelberg, Dordrecht, London, pp 35-72 (Chapter 3). DOI: 10.1007/978-1-4614-4960-7 ISBN 978-1-4614-4959-1

11   Goulson, D. and Williams, P. 2001. *Bombus hypnorum* (Hymenoptera: Apidae), a new British bumblebee? *British Journal of Entomology and Natural History* 14: 129–131

12   https://www.bumblebeeconservation.org/tree-bumblebee-*bombus*-hypnorum/ Accessed December 2022.

13   Velthuis and van Doorn 2006. See Chapter 4 note 4.

14   Ings, T.C., Ward, N.L. and Chittka, L. 2006. Can commercially imported bumble bees out-compete their native conspecifics? *Journal of Applied Ecology* 43: 940-948. DOI: 10.1111/j.1365-2664.2006.01199.x

15   Natural England, UK government. "Licence to permit the release of non-native sub species of bumblebee (*Bombus terrestris*) in commercial glasshouses for research." https://www.gov.uk/government/publications/bumblebees-licence-to-release-them-for-pollination-and-research/licence-to-permit-the-release-of-non-native-sub-species-of-bumblebee-*bombus*-terrestris-in-commercial-glass-houses-for-research Accessed January 2023.

16   Agriculture and Horticulture Development Board. 2022. *Tomato: Phase 3 of an investigation into poor pollination performance by the native bumblebee, Bombus terrestris audax*. British Tomato growers association final report, 16 March 2022 https://horticulture.ahdb.org.uk/pe031b-tomato-phase-3-of-an-investigation-into-poor-pollination-performance-by-the-native-bumblebee-bombus-terrestris-*audax*

17   Velthuis and van Doorn 2006. See Chapter 4 note 4.

18   Goulson, D. 2013. *A Sting in the Tail*. Jonathan Cape, London, UK. 256 pp.

19   Sladen 1912. See Chapter 1, note 4.

20   https://en.wikipedia.org/wiki/*Bombus*_polaris Accessed January 2023.

21   Gurr, L. 1957. Seasonal availability of food and its influence on the local abundance of species of bumble bees in the South Island of New Zealand. *New Zealand Journal of Science and Technology*. 38A: 867 – 870

22   Goulson 2012. See note 1.

23   Evans, E,. Thorp, R., Jepsen, S., and Hoffman Black, S. Status Review of Three Formerly Common Species of Bumble Bee in the Subgenus *Bombus*. *The Xerces Society*. https://xerces.org/publications/scientific-reports/status-review-of-three-formerly-common-species-of-bumble-bee Undated. Retrieved January 2023.

24   Cameron, S.A., Lorzier, J.D., Strange, J.P., Koch, J.B., Cordes, N., Solter, L.F. and Griswold, T.l. 2011. Patterns of widespread decline in North American bumble bees. *PNAS* 108(2): 662-667. DOI: 10.1073/pnas.1014743108

Szabo, N.D., Colla, S.R., Wagner, D.L., Gall, L.F. and Kerr, J.T. 2012. Do pathogen spillover, pesticide use, or habitat loss explain recent North American bumblebee declines?. *Conservation Letters* 5: 232-239. DOI: 10.1111/j.1755-263X.2012.00234.x

Soroye. P., Newbold, T. and Kerr, J. 2020. Climate change contributes to widespread declines among bumble bees across continents. *Science* 367: 685–688. DOI: 10.1126/science.aax8591

25   Tamburini, G., Pereira-Peixoto, M-H., Borth, J., Lotz, S., Wintermantel, D., Allan, M.J., Dean, R., Schwarz, J.M., Knauer, A., Albrecht, M. and Klein, A-M. 2021. Fungicide and insecticide exposure adversely impacts bumblebees and pollination services under semi-field conditions, *Environment International* 157:106813. DOI: 10.1016/j.envint.2021.106813.

26   United States Environmental Protection Agency. Summary of the Endangered Species Act. https://www.epa.gov/laws-regulations/summary-endangered-species-act Accessed January 2023.

27   U.S. Fish and Widlife Service. 2019. *Draft recovery plan for the rusty patched bumble bee (Bombus affinis).* Midwest Regional Office, Bloomington, Minnesota. USA. 10 pp. https://www.fws.gov/sites/default/files/documents/Final%20 Recovery%20Plan%20_Rusty%20 atched%20Bumble%20Bee_2021.pdf

28   Smith, T.A., Strange, J.P., Evans, C, Sadd, B.M, Steiner, J.C., Mola, J.M and Traylor-Holzer, K. (Eds.). 2020. *Rusty Patched Bumble Bee, Bombus affinis, Ex Situ Assessment and Planning Workshop: Final Report.* IUCN SSC Conservation Planning Specialist Group, Apple Valley, MN, USA.

29   https://wildlifepreservation.ca/bumble-bee-recovery, accessed January 2023.

30   Mallinger, R.E. and Gratton, C. 2015. Species richness of wild bees, but not the use of managed honeybees, increases fruit set of a pollinator-dependent crop. *Journal of Applied Ecology* 52: 323–330. DOI: 10.1111/1365-2664.12377

      McGrady, C.M., Strange, J.P., López-Uribe, M.M. and Fleischer, S.J. 2021. Wild bumble bee colony abundance, scaled by field size, predicts pollination services. *Ecosphere* 12(9): e03735. DOI 10.1002/ecs2.3735

31   Lemanski, N, Williams, N.M. and R Winfree R. 2022. Greater bee diversity is needed to maintain crop pollination over time. *Nature Ecology and Evolution* 2022 Oct;6(10):1516-1523. DOI: 10.1038/s41559-022-01847-3.

32   Kleijn, D., Winfree, R., Bartomeus, I. *et al.* (60 co-authors). 2015. Delivery of crop pollination services is an insufficient argument for wild pollinator conservation. *Nature Communications* 6: 7414. DOI: 10.1038/ncomms8414

33   UN Convention on Biological Diversity. 2008. *Agricultural biodiversity // why is it important?* https://www.cbd.int/agro/importance.shtml Accessed February 2023.

34   SARE 2012 *Maintaining Companion Planting Techniques while Mechanizing in Diverse, Small-Farm Vegetable Operations.* Sustainable Agriculture Research & Education. https://projects.sare.org/sare_project/fnc10-814/ Accessed February 2023.

35   OECD data on governments' support for agriculture: https://data.oecd.org/ agrpolicy/agricultural-support.htm Accessed January 2023.

36 Morales, C.L., Arbetman, M.P., Cameron, S.A. and Aizen, M.A. Rapid ecological replacement of a native bumble bee by invasive species. *Frontiers in Ecology and the Environment.* 11(10): 529-534. DOI:10.1890/120321

37 Nagamitsu T., Yamagishi H., Kenta T., Inari N. & Kato E. 2010. Competitive effects of the exotic *Bombus terrestris* on native bumble bees revealed by a field removal experiment. *Population Ecology* 52: 123-136.

38 One-third of workers foraging: Roland Kreskóci, Koppert, personal communication, and a recent count I did with trapping vestibules on observation hives of *B. terrestris*: twenty foragers from a total number of sixty.

39 Tod 1986. See Chapter 4, note 9.

40 Tasei, J-N., and Aupinel, P. 2008. Validation of a Method Using Queenless *Bombus terrestris* Micro-Colonies for Testing the Nutritive Value of Commercial Pollen Mixes by Comparison with Queenright Colonies. *Journal of Economic Entomology.* 101: 1737–1742. DOI: 10.1603/0022-0493-101.6.1737

41 Strange, J.P., Raising Bumble Bees at Home: A Guide to Getting Started. https://www.ars.usda.gov/ARSUserFiles/20800500/BumbleBeeRearingGuide.pdf

Owen, R.E. 2016 "Rearing Bumble Bees for Research and Profit: Practical and Ethical Considerations," in E. D.Chambo (ed.), Beekeeping and Bee Conservation - Advances in Research. (Chapter 9) pp 225-242. DOI: 10.5772/63048

42 For information on the BumbleBox™ please email bumblebox@plantandfood.co.nz. The New Zealand Institute for Plant and Food Research Limited is a New Zealand government-owned Crown Research Institute.

43 Hiraguri, T., Kimura, T., Endo, K. *et al.* 2023. Shape classification technology of pollinated tomato flowers for robotic implementation. *Nature, Scientific Reports.* 13: 2159. DOI: 10.1038/s41598-023-27971-z

44 Robinson, J. 2020. *Steve Jobs, typographer: An essay on the typographic legacy of Apple's founder.* https://uxplanet.org/steve-jobs-typographer-2e450a356437 Accessed February 2023.

45 Mader, E., Spivak, M. and Evans, E. (eds) 2010. *Managing Alternative Pollinators: A Handbook for Beekeepers, Growers, and Conservationists, SARE Handbook 11.*, SARE (Sustainable Agricultre, Outreach And Education) and NRAES (Natural Resource, Agriculture, and Engineering Service). https://www.sare.org/wp-content/uploads/Managing-Alternative-Pollinators.pdf

Appendix 2. Bumblebee genetics and diploid males

1   Murdoch, W.W., Nisbet, R.M., Luck, R.F., Godfray, H.C.J., and Gurney, W.S.C. 1992. Size-Selective Sex-Allocation and Host Feeding in a Parasitoid--Host Model. *Journal of Animal Ecology*. 61(3): 533–541. DOI 10.2307/5608

2   Crozier R.H. 1971. Heterozygosity and sex determination in haplo-diploidy. *American Naturalist* 105: 399-41.

Duchateau, M.J., Hoshiba, H. and Velthuis, H.H.W. 1994. Diploid males in the bumble bee *Bombus terrestris. Entomologia Experimentalis et Applicata* 71: 263-269. DOI: 10.1111/j.1570-7458.1994.tb01793.x

Nelson Pomeroy **Bumblebee Keeper**

www.ingramcontent.com/pod-product-compliance
Lightning Source LLC
Chambersburg PA
CBHW050037220326
41599CB00040B/7196